D0769305

Due

Gauss

Carl Friedrich Gauss

W. K. Bühler

Gauss
A Biographical Study

With 10 Illustrations

Springer-Verlag
Berlin Heidelberg New York

AMS Subject Classification (1980): 01A55, 01A70

Library of Congress Cataloging in Publication Data

Bühler, Walter Kaufmann-, 1944–
  Gauss: a biographical study

  Bibliography: p.
  Includes index.
  1. Gauss, Karl Friedrich, 1777–1855.
2. Mathematicians—Germany—Biography.
QA29.G3B83   510'.92'4 [B]   80-29515

Frontispiece: Gauss in 1803, after a portrait by J.C.A. Schwartz

9 8 7 6 5 4 3 2 1

ISBN 3-540-10662-6 Springer-Verlag Berlin Heidelberg New York
ISBN 0-387-10662-6 Springer-Verlag New York Heidelberg Berlin

Preface

Procreare iucundum,
sed parturire molestum.
(Gauss, sec. Eisenstein)

The plan of this book was first conceived eight years ago. The manuscript developed slowly through several versions until it attained its present form in 1979. It would be inappropriate to list the names of all the friends and advisors with whom I discussed my various drafts but I should like to mention the name of Mr. Gary Cornell who, besides discussing with me numerous details of the manuscript, revised it stylistically.

There is much interest among mathematicians to know more about Gauss's life, and the generous help I received has certainly more to do with this than with any individual, positive or negative, aspect of my manuscript. Any mistakes, errors of judgement, or other inadequacies are, of course, the author's responsibility. The most incisive and, in a way, easiest decisions I had to make were those of personal taste in the choice and treatment of topics. Much had to be omitted or could only be discussed in a cursory way.

Nicolaus von Fuss, a man not unknown to the student of Gauss's life, begins his "Lobrede auf Herrn Euler" (Euler, Opera I) with a wonderful description of the biographer's task:

To describe the life of a great man who distinguishes his century by a considerable degree of enlightenment is to eulogize the human mind.*

This biography is an attempt to follow Fuss's program, even though I do not know whether Gauss would have subscribed to his enlightened statement. He would not, I hope, have objected to it in connection with the lives of the very great ones, Archimedes and Newton.

The book contains many quotations, even lengthy passages, from Gauss's writings. Originally, I intended to quote in the original languages but was

---

*Fuss continues with an extensive list of requirements which a biographer should satisfy. I cannot disagree with him; nor can I claim to know as much as Fuss demands.

persuaded to translate these passages into English. Stylistically this is no gain, because Gauss's German is clear and very energetic—*kraftvoll*, as one might say in German. The originals of nearly all the quotations can be found in the "Notes" section at the end of the book.

Finally, I have the pleasant duty of thanking several libraries whose excellent services I was able to use when writing this book: the library of the Courant Institute of Mathematical Sciences of New York University, the Public Library of the City of New York, and the Niedersächsische Staatsbibliothek in Göttingen, the depository of Gauss's Nachlass.

This book is dedicated to my wife who never seemed to be jealous of her older rival.

March 1981   W. K. Bühler
Mt. Kisco, New York

Contents

Introduction

The year 1977 was the bicentennial of Gauss's birth; 1980 marks the 125th anniversary of his death. Thus, Gauss was formed by experiences that go very far back, to the now remote world of the 18th and the first decades of the 19th century. These times can no longer be understood easily.

Every year, it will become more complicated to write a biography of Gauss. More and more extraneous material has to be supplied and to be explained to a reader whose interest may not be primarily historical. Right now, we appear to have reached a critical juncture—in a few years, the world will be very different, and the vestiges of the age of Enlightenment and the romantic age will no longer be recognized as part of our heritage.

This book describes the life of the mathematician and scientist Carl Friedrich Gauss. Gauss lived in a period of extraordinary political and social developments, even measured by the standards of our own fast moving and eventful age. He was 12 years old when the French Revolution broke out, 29 when the seemingly eternal, 1000-year-old Holy Roman Empire was dissolved, 38 when Napoleon was defeated, and over 70 when Germany had its own liberal revolution in 1848. During the same period, the so-called first industrial revolution took place, with its lasting and incisive effects on everyday life and the political and social order. It is clear that all this affected Gauss's life in an explicit and tangible way. There were obvious moments like the realization of formerly impossible large-scale and efficient scientific experiments or the improvement of telescopes and other optical instruments, but we will also encounter more subtle effects like the consequences of the replacement of the old feudal order by "new style" absolutist governments in Germany and the increasing luxuries of everyday life.

This biography is addressed to the contemporary mathematician and scientist, not to the historian of science or the psychologist collecting the scalps of great men. Its aim is modest—this is not an attempt to write a definitive "Life of Gauss." At the same time, it is immodest if not grandiose to try to select from Gauss's life and work those facets which are of contemporary interest and palatable to a reader who is not primarily historically motivated nor has any immediate reasons why he should be.

Gauss's work, particularly his mathematics, has survived and is of scientific interest even today. It is worth studying not only for curiosity's sake but also as a deep source of inspiration. Gauss had much insight into matters which even today we have not fully understood. Good scientific work, particularly in mathematics, has an ageless quality. This is the inner reason for our curiosity about Gauss; it is neither pointless nor futile to try to understand his ideas and his life. This biography was conceived as a guidebook to an area which, though difficult to enter and not well charted, is rich in rewarding results.

In order to be useful, a guidebook must be selective. The reader, though not completely at his guide's mercy, has to have confidence in his knowledge and, more importantly, his tastes. Whenever there was a choice in this book, it was against generalities: specific examples are used to show Gauss's way of thinking, to illustrate his work, and to summarize in what way it differs from that of his predecessors and contemporaries. The decision what to include or omit was often arbitrary, and many important results could not be mentioned. We hope the global picture is satisfactory even if there are many inconsistencies, omissions, and mistakes in detail.

Gauss's personal fate, his uninterrupted development in nearly always benevolent and helpful surroundings, was quite exceptional for Germany. This is of a certain interest to the historian but must also be kept in mind by the student who is not primarily concerned with the historical aspects of Gauss's life. The absence of obstacles influenced Gauss's actions at several points of his career and development; moreover, it is a noticeable factor in the way he thought. That Gauss does not "fit" into the German intellectual scene of the 18th or 19th century is clear; that his way was so exceptionally smooth may be one of the reasons why the older secondary literature shows so little understanding for many of the nonscientific aspects of Gauss's life.

Of the secondary sources, the two most important and useful ones are both connected with the name of Felix Klein. One is Klein's summary of Gauss's work in his lectures on the development of mathematics in the 19th century, the other the collection of essays which, following Klein's initiative, were included in Volumes X and XI of the Academy Edition of Gauss's collected works. Klein's work and the essays are discussed in the Epilogue and in one of the appendices of this book. Originally, the collected works were supposed to contain a biography of Gauss, but this part of Klein's plan never materialized, nor did the comprehensive index that was also scheduled to appear.

There is very little in this biography that is not already known to the specialist. Practically all the information which was used can be found in published and printed sources. The collected works are reasonably complete; most of Gauss's relevant correspondence (and a lot that is irrelevant) has been published. There is more about the primary and secondary sources in two of the appendices at the end of the book.

This biography could never have been written without the vast secondary literature that has been devoted to Gauss's life and work (cf. in this connection my remark at the head of the Bibliography). Klein's work and the essays in Vols. X and XI of the collected works have already been mentioned; another indispensable source is G. W. Dunnington's biography *C. F. Gauss— Titan of Science* (1955). It contains an enormous wealth of material, the result of several decades' labor.

Due to the character of this biography it was difficult to assess how much substantiation or how many details a reader would expect or want to know. The various footnotes which accompany the text provide the interested reader with references and, occasionally, additional information. They can be skipped by readers who are in a hurry or of a credulous disposition. Naturally, much of what is said had to go without substantiation, but I took care to qualify most of my more dubious assertions.

CHAPTER 1

Childhood and Youth, 1777–1795

Only a very few characteristic and interesting facts are known from the childhood and youth of Gauss. Basically, we have to satisfy ourselves with the bare data of his biography and that kind of information which can be induced from a general familiarity with the social and political situation of the time. Our only major specific source is Gauss himself; the stories of his childhood that he liked to tell as an old man were preserved and transmitted by his students and friends[1] and are now part of our traditional picture of Gauss. Many of these anecdotes cannot be substantiated, nor are they of serious interest.

Later investigations have unearthed some details about the origins of Gauss's family and the fate of some of his relatives. The culmination of these efforts is a vast family tree which extends into such areas as the American midwest and the borough of Brooklyn; in addition to Gauss's direct relatives and descendants it includes a number of far-removed and even questionable, only potential, branches of the family. In Germany, not many direct descendants of Gauss have survived, but the family seems to be flourishing in the United States.[2]

Johann Friedrich Carl Gauss was born on April 30, 1777. He was the only known offspring of the marriage between Gebhard Dietrich Gauss, born in 1744, and Dorothea Benze, who was one year older than her husband. There was another boy, a few years older, from a previous marriage of the father.[3] Gauss's paternal family, small farmers originally, had moved to the city of Brunswick around the year 1740, exchanging the life of farmhands or virtually rightless tenants for the equally poor, but at least hopeful, existence of "half-citizens" in Brunswick, the main city of the Duchy of Brunswick-Wolfenbüttel. As to many other newcomers of this period, the move offered to the Gauss family an expectation of slow prosperity and the promise of a brighter future. There was no easy way to improve one's lot; the medieval guilds (*Zünfte*), by controlling the trades, controlled much of the life of the city and did not permit any economic expansion. They were closed to newcomers, and even Gauss's father, one generation later, could not gain access to them and had to earn his living in a sequence of menial and un-profitable jobs, as a gardener, a canal worker, a street butcher, and an accountant for a funeral society.[4]

One of the family's first important projects was to acquire their own house. Only if one possessed a house within the city limits could one become a full citizen with all the accompanying rights and privileges.[5]

A few years after the house was acquired, the world into which Carl Friedrich had been born ceased to exist. It was a sudden and unexpected catastrophe when the German states, among them Brunswick, were overrun by the victorious armies of revolutionary France. The 1780s, as Gauss was growing up, did not offer any vision of a different or better future.

The birthplace of Carl Friedrich Gauss was a small house in a street called "Wendengraben". Later, the address of the house was changed to "Wilhelmstrasse 30".[6] A few years after their son was born the parents moved, leaving the scene of one of Gauss's best-known childhood stories: supposedly, the future prince of mathematicians almost drowned in a nearby canal when he was three or four years old.

Neither of Gauss's parents had much education: his father, as we can gather from his jobs, at least could read and write; he also knew elementary arithmetic. His mother probably could read, but not write.* Gauss does not seem to have been close to his father, and we know from later remarks that he traced his genius back to his mother. In 1817, after her husband's death, she moved to Göttingen to live in her son's household until she died in 1839, at the age of 96. Dorothea Gauss was the offspring of a family of stonemasons and had moved to Brunswick in 1769; the origins of her family, like those of her husband's, are to be found to the north of the city. Carl Friedrich's first wife appears to have been among her acquaintances, perhaps from the time before her marriage when she seems to have worked as a maid in the household of the Ritter family. We can assume that his mother was Carl Friedrich's main source for stories of his childhood[7] and for what he knew of the extended family.

The factual information that is available to us starts with the year 1784 when the young Carl Friedrich entered elementary school. That he actually went to school was not unusual at the time, for children who grew up in a major city usually had this opportunity.[8] He was unusually lucky in another way, for his teacher, Büttner, seems to have been competent and concerned. He took a personal interest in the boy, trying to help and encourage him. In retrospect, it is easy to see how the bright youngster must have excelled among the other students. Nevertheless, he had to be recognized from among

---

* A letter from Gauss to his future second wife (quoted on p. 68) gives some additional infor-mation. We know that Dorothea Gauss could not read handwriting in 1810 but this may have been an early symptom of her later blindness. To infer more is pure speculation.

more than 50 children of various ages and levels of knowledge who sat with Gauss in the same classroom. Gauss reportedly knew how to read and write before entering school, skills he appears to have picked up without any help from his parents.[9] When he was barely three years old he could count and perform elementary calculations. Büttner was impressed by the boy; in 1786 he obtained from Hamburg a special arithmetic text for his exceptional student—there was nothing in the standard primers which would have been new to him. Büttner's assistant during these years was Martin Bartels (1769–1836), later professor of mathematics at the University of Kazan and only eight years older than Gauss. He soon recognized Gauss's genius* and devoted special attention to the young Carl Friedrich. We do not know how the parents encouraged their prodigious son, if they did so at all; the times, against a background of need and poverty, were utilitarian and not conducive to appreciating the advantages of good schooling and academic success. The parents' benevolent but disinterested astonishment at their son's achievements was most certainly not coupled with any expectation of an extraordinary career. In their narrow and limited world, after all, there were more important and promising gifts for a son of a dependent laborer to have than a curious facility in counting and arithmetic.

Carl Friedrich left the small world of his parents in 1788 when Büttner helped him to be admitted to secondary school. The lectures at the new school were orderly and regular; for the first time, Gauss studied in reasonably sized classes, along with other students of a similar level of knowledge and age. Even according to the standards of the time, the curriculum was old-fashioned and imbalanced, with a bizarre overemphasis on the ancient languages, particularly Latin.

Thus, most of what proved to be decisive to Gauss's further development had happened by the end of the 1780s, well before the old political and social order broke down. Gauss was 12 years old at the beginning of the French Revolution and nearly 30 when its effects were felt in Germany. Only after 1806 did society become more democratic; by this time, most of Gauss's personal social development was complete. His horizons were certainly narrow; from his own perspective as well as his parents', his early successes did not promise much. Only the modest and tenuous affluence that his father and his half-brother, a weaver, hoped for could be conceived. Perhaps the ultimate achievement would have been the questionably attractive career of a Protestant parson or teacher.

Gauss used his time at the new school well, acquiring a solid knowledge of Latin, the indispensable prerequisite for the pursuit of higher learning and an academic career. He also learned to use the official High German, the language of Luther's translation of the Bible. Up to then, Gauss had only spoken the local dialect.[10]

---

* As one would then have said. See the discussion of this word below (p. 14).

In 1791, Carl Friedrich, being a gifted and promising young citizen, was introduced to his prince, the Duke of Brunswick-Wolfenbüttel. The Duke was apparently impressed and granted Gauss an initial yearly stipend of 10 talers. Awards like this were not unusual for the time, especially in small and well-administered states like Brunswick. They should be seen as fore-runners of today's official, regularized scholarships. Without them, the so-cial barriers would have been impenetrable for a gifted boy; they were an important factor in the development of an efficient and loyal civil service, an essential tool for the absolutist governments of the time.

Gauss would never have advanced without direct help from a number of people interested in promoting his talents. Most important was the help of the *Hofrath* (councillor) von Zimmermann, a professor at the academy Collegium Carolinum, a high civil servant, and personal factotum of the Duke—a typical representative of an absolutist administration. His benev-olent influence accompanied Gauss until 1806, the year in which the state of Brunswick-Wolfenbüttel was broken up by Napoleon, and one year before Gauss became director of the observatory in Göttingen. There are many instances where Gauss shows his gratitude and the high esteem in which he held Zimmermann; examples can be found in his correspondence with the astronomer Olbers, who was later himself to act in a similar role on Gauss's behalf: as early as 1802, Olbers suggested to the Hanoverian govern-ment that Gauss be made director of the observatory in Göttingen.

Selected because of his good academic record, Gauss spent the years 1792–1795 as a student at the Collegium Carolinum, a new academy, barely ten years old. It was a progressive, science-oriented institution, one of the best schools of its kind, founded and run directly by the government, whose principal source it was for well-qualified, loyal bureaucrats and military personnel. Academies of this type were not rare in Germany. By nature they were elitist, and many of the better-known writers and scientists of the 18th and early 19th centuries received their education at them.[11]

Gauss's home could no longer contribute to his development and intel-lectual progress; what Gauss knew, he owed to his teachers and his own genius. Gauss does not appear to have minded this independence, he does not even seem ever to have been emotionally very close to his parents. We only know his attitude as an adult; by then, he was very detached: for example, he depicted his father as honest and hard-working but also as narrow-minded and irascible, and he mentioned, in a surprisingly open way, that the temperaments of his parents had not been compatible and their marriage had not been happy.*[12] To be given the opportunity to increase his knowledge and exercise his mind must have meant everything to Gauss

---

* This sentence should not be overinterpreted because Gauss did not fully share our concept of (and devotion to) the family. Our attitudes are very much a product of the 19th century. Gauss adopted the modern concept of family later in his life, but we should not assume that he readily applied it to his parents.

at that time. The school was at the center of his life, and nonacademic interests and demands at the periphery.

When speculating about Carl Friedrich's success at school we should keep in mind that his road would have been incomparably more difficult had he not been a good student of Latin and Greek. Without this, his progress could never have been as orderly as it was, regardless of his mathematical ingenuity and achievements. There are enough contemporary examples of how the rigidity of formal education and the overwhelming importance given to the classical languages broke the careers of promising young men. Among those who could not cope with the system are men such as von Basedow, Wenzel, and Arndt, who occasionally are referred to as *Schreckensmänner* (horror men).[13]

During his years at the Carolinum, Gauss met several young men with whom he formed lasting friendships, among them A. W. Eschenburg and K. Ide. Eschenburg's father was a teacher at the academy whose translations of several of Shakespeare's plays had been acclaimed by C. M. Wielandt, one of Germany's most famous poets. His son, Gauss's lifelong friend, later became a highly placed civil servant in the Prussian administration. Ide, like Gauss, became a mathematician and astronomer. He accepted a position at one of the Russian universities and died young. Yet another friend was Meyerhoff who, not much later, was to correct and polish Gauss's Latin in *Disquisitiones Arithmeticae*.

The library of the Collegium Carolinum was unusually good and one can assume that it contained much of the classical mathematical literature. Gauss used the time at the Carolinum to study many of these texts, among them Newton whom he much admired. (He and Archimedes were the only ones to receive the attribute "illustrissimus" from Gauss in his citations). Other important books were Euler's Algebra and Analysis and several of Lagrange's works.*

When he left the Collegium Carolinum, Gauss knew enough to understand the current literature. He was able to do independent research without rediscovering an inordinate number of known results. From his diary we know that Gauss was mostly concerned with number theory and algebra at the time, but it would be wrong to make rigid distinctions between the different fields of mathematics at this stage of Gauss's work. The time at

---

* We cannot be completely certain what books were accessible to Gauss at the time. In his *Gauss zum Gedächtnis*, Sartorius explicitly denies that Gauss had any access to advanced books before coming to Göttingen, but there are reasons to disagree with Sartorius. There are later statements by Gauss himself; more importantly, the level and maturity of his earliest mathematical writings seem to support our assumption. Substantial indirect evidence can be gained from the list of the books which Gauss borrowed from the Göttingen university library as a student (reprinted in [Dunnington]). It contains the proceedings of the St. Petersburg Academy and other research literature, but very few of the classical texts or treatises with which Gauss was presumably familiar.

the Carolineum was essentially one of receptivity; Gauss's hunger for mathematics was universal and omnivorous, though it expressed itself, as is usual, in specific problems and questions. His good but unsystematic knowledge of the older literature, acquired in this period, is reflected in his historical remarks and in the seemingly abrupt switches between topics which are conspicuous in the apodictic assertions of the diary.

We have seen how Gauss's genius was already recognized in elementary school; presumably the best contribution his teachers made to his education was to give him the opportunity of studying independently. His first spectacular success came when he was not quite 19 years old—the proof of the constructibility of the regular 17-gon with straightedge and compass. It is often claimed that this discovery prompted Gauss to devote his academic activity to mathematics rather than classical languages. Zimmermann announced the result for Gauss in the journal *Intellegenzblatt der allgemeinen Litteraturzeitung.*[14]

Other questions with which Gauss concerned hinself during these years (and as early as 1791) were the (elementarily accessible, but complicated and far-reaching) theory of the arithmetico-geometric mean and the distribution of primes. Both lead to interesting numerical calculations. Gauss's interest in numbers and their manipulation was strong and not confined to his early years.

The Contemporary Political and Social Situation

This short digression will try to make up for our insufficient knowledge of details from Gauss's childhood and youth by providing some general comments about the social and historical situation of the time. Gauss grew up in Brunswick, a city located in the (Protestant) north of Germany. Though religion was still an important facet of everyday life, his parents do not seem to have been very religious and were certainly free of the evangelical (pietistic) tendencies which were then widespread among families of a similar back-ground.[1] The sudden interest, during the 1840s, in experiments like seances and table-turning may have reminded Gauss of the religious folklore of his youth. His comments were predictably and decidedly negative when he was consulted by W. Gerling, a former student who was professor of physics at the University of Marburg. Gauss completely rejected mystical experiences of this (or any) kind and made it very clear to Gerling that there was no scientific basis.[2]

Brunswick was an old and once wealthy center of trade. It had been a member of the medieval Hanseatic League but had seen a steady decline for the last 150 years. In the middle ages and even as late as the early years of the 17th century, Brunswick was an important commercial city, a serious rival of Hamburg or Amsterdam. It was politically independent though nomi-nally under the suzerainty of the duke of Brunswick-Wolfenbüttel. Governed according to an old oligarchic constitution by an elected council whose members habitually came from the same families, Brunswick was weakened by popular revolts and the disastrous consequences of the Thirty Years War. It lost its political independence without a fight in 1671 when it was incor-porated into the duchy of Brunswick-Wolfenbüttel, whose capital it became in 1736. As a result of these revolutions, many of the old patrician families left Brunswick for Hamburg and the Low Countries, leaving the city and its economy to deteriorate even more, until it reached a low point around 1750, when it counted only 20,000 souls. Slowly these losses were made up by the influx of immigrants from the vicinity, who were now able to move into the capital quite freely. This group included Gauss's paternal family, which came from an area to the north of the city.

In Brunswick-Wolfenbüttel, as in the other small German states of that time, agriculture was the backbone of the economy, providing the financial basis for the government. Several attempts to industrialize, following the French and Prussian examples, were not too successful. This is not surprising, given that "industrialization" meant projects like the systematic cultivation of mulberry trees and breeding of silkworms, or the canalization of the River Oker, an unrealistic project that was dropped after some initial feeble attempts towards its realization.

Throughout Germany the big cities, lifeless monuments to past glories, had turned into gigantic almshouses. The rigid old social order did not seem to waver or show any sign of the germinating radical changes and momentous revolutions destined to make the city once more the center of progress and civilization. One fact, not in itself of much importance, will show the stifling climate of the time, just before the great storm: several imperial edicts, issued to protect the well-organized clothmakers, forbade the use of the mechanical loom, the same machine whose success marked the start of the industrial revolution in England.

Public education was one of the few areas of genuine progress in the 18th century; here the groundwork was laid for the changes which took place in the 19th century. Though compulsory schooling was not strictly enforced, the majority of Gauss's generation could read, write, and do elementary arithmetic. In some districts, even elementary Latin was taught. Gauss's own career is a good example of the advancement of a gifted child of humble background—there was no hope of advancing without the help and personal interest of a prince, a rich merchant, or some other wealthy and influential sponsor. The Collegium Carolinum was only one among a number of similar schools; more famous examples are the academy in Schulpforta and the Karlsschule in Stuttgart. With the emergence of these schools, the churches lost another of their traditional strongholds in education, having already, at least in most of the Protestant countries, lost much of their control over the universities during the Reformation.[3]

The novel *Anton Reiser*,[4] less fiction than a bitter and realistic account of the author's childhood and adolescence, gives a good idea of what it meant to grow up as a poor and gifted child in the Protestant north of Germany in the second half of the 18th century. The author, C. Moritz, came from a background similar to Gauss's; the book describes the limitations, deprivations, and humiliations of poverty and the emotional and intellectual struggle of a child growing up under the protection of an unknown benefactor.* It was customary and natural that the direct protection of an alumnus like Gauss ended when the young man left the gymnasium. This was another

---

* Like its author, the protagonist of the novel had difficulties finishing school and was not able to pursue his education and career in the expected orderly and detached fashion. Had he not been able to express himself as a writer, Moritz would have died penniless and would, like so many others, not be remembered. The book is an excellent source for the social and cultural background of Gauss's youth, especially, as Moritz grew up in the city of Hanover, only some 30 miles west of Brunswick.

critical point, often leading to the sudden and premature end of a promising career. Not, as we shall see, in the case of Gauss—his path was straight and orderly, without the nearly inevitable surprises and reverses.

When we contemplate the life of a man, we see it through the psychological and sociological theories which have grown up over the last 75 years. We have learnt that a man's parents, his relationship with them, and, in general, social and psychological factors, are of decisive importance. We lack all these essential details for a discussion of Gauss's life and have to ask ourselves whether our task must not then be hopeless. We may indeed approach it with empty hands, but need not be too unhappy about our lack of knowledge. It is conceivable that the psychological and sociological approach is not simply the proud result of the scientific progress of the last hundred years; it may just as well be a symptom of these years. One might well assume that the new categories which we have learnt to apply since the end of the last century are not so suitable for the understanding of an earlier period, even one as close to us as Gauss's. We will consequently not be too concerned that such information is not available: at least we shall be protected from misunderstandings and our (presumably innate) tendencies towards modernistic interpretations.

Though our approach will in general be "nonrevisionist" we will enter into a detailed psychological discussion of Gauss's family life. Gauss's relations with his two wives and with his children from both marriages deeply affected his development as a man and as a scientist. We shall see how his *Familiensinn*, his protective interest in the social advancement of his family and in the careers of his children, developed well in line with the general tendencies of the time. Gauss's concern was radically different from what he had experienced, growing up in the 18th century world of his parents; only in retrospective interpretations has his attitude been identified with the middle class ideal of family, one of the first products of the early 19th century. Gauss's contemporaries and friends, the Schumachers, Olbers, and Bessels, but particularly his second wife, may have influenced him. The clue was quickly taken up by his younger contemporaries and biographers. Private, individual explanations for certain actions of Gauss were readily provided even if other ("objective") motives could have been found. An example is the reason traditionally cited for the fall-out with the astronomer Bessel in 1832: Gauss supposedly did not write sufficiently many letters of recommendation for Bessel's son-in-law, a geographer; equally Bessel is usually faulted for not sending a letter of condolence when Gauss's second wife died.[5] We shall return to this conflict below, because Bessel was, with the possible exception of Olbers, the only regular correspondent who was a competent mathematician and an equal partner as a theoretical and observing astronomer.

Early in his adult life, Gauss severed most of the "meaningful" social and emotional ties that a man could have. He was not close to his parents, nor do we know of other strong attachments from his childhood which survived into his adult life.

We now turn to a question which appears to be awkward but is perfectly normal when interpreted in the terms of the 18th century, when the word "genius" had a special meaning. Gauss had no interest in appearing to be a genius and becoming part of a movement which was very much in vogue when he was young. There was a regular cult—not only among artists—of the young, gifted, and creative; one symptom of creativity and genius was a disregard for established rules in one's personal life and creations. Such a concept was quite original and revolutionary for the time, but despite its nationalistic and anti-French twist, Gauss did not share it or betray any sympathy for it. Classical education, whether of the ancient or of the French variety, was rejected, together with the uncritical confidence in the forces of progress that was so typical of the Age of Enlightenment. The genius—a true, unspoilt product of Nature—created his own rules and lived, worked, and developed his personality in accordance with them. In Germany, the artists of the *Sturm und Drang* era were the main proponents of this movement. They in turn inspired many of the generally better-known poets and philosophers of the romantic school, among them F. Hölderlin and G. W. Hegel.[6] There will be incidents in Gauss's later life which will let us understand his complete rejection of this romantic attitude; his childhood and youth did nothing to make Gauss more sympathetic to the world of the noble savage.

Sturm und Drang is the German variety of the universal European intellectual unrest at the end of the 18th century. Its specific roots have to be sought in the political situation of the time, particularly in the desire of the rising middle class for a constructive role in society. The optimistic creed of the Enlightenment appeared to the impotent citizen to be as irrelevant as the sterile orthodox dogmatism which had preceded it. The majority of the *Sturm und Drang* and later of the romantic poets belonged to the same social group as supported and sustained the great revolution in France, but in less-developed Germany a political and social movement could never hope to be politically successful without assistance from abroad. Gauss lived in a different world, one that was not affected by these commotions. For at least the first thirty years of his life, Gauss felt no conflict between his individual development and the feudal political order. From his own perspective, his advancement appeared to be the best example of the way the enlightened absolutist variety of feudalism and its political and pedagogical theories worked.

Yet another, more speculative inference can be made. Not only his background and upbringing, but also his experiences as a scientist and mathematician may have exerted a conservative influence on Gauss. It is not possible to quantify this statement in a meaningful way, but now and again Gauss's life will supply us with clues which indicate how upset he was by any changes that interfered with his scientific work or even his daily life.[7]

CHAPTER 2

Student Years in Göttingen, 1795–1798

In 1795, at the age of eighteen, Gauss left his native Brunswick to study mathematics at the University of Göttingen, a small town approximately 65 miles to the south. Göttingen, in the state of Hanover, was already "abroad"; Gauss went there against the wish of his Duke, who wanted him to attend the local university of Helmstedt.* Helmstedt was an old school, without much scientific reputation, dominated as it was by its schools of divinity and law. We know that Gauss preferred Göttingen because of its good library,[1] but its reputation as a science-oriented "reform" university may also have contributed to his decision. Göttingen university, founded by King George II of England (who was also Prince of Hanover) after the pattern of Oxford and Cambridge, was better endowed than most other German universities and also more independent of governmental and ecclesiastical supervision and interference. There was not even a school of divinity; instead, the schools of medicine and the natural sciences were cultivated.[2] As was customary at the German universities, Gauss was completely on his own as a student, a beneficiary of "academic freedom". There were no regulations as to what lectures to attend, no tutors who worked with him, no examinations or curricular control, not even within the community of the students.[†] Gauss had social contacts with several of the professors, among them the physicist Lichtenberg, the astronomer Seyffer, and the linguist Heyne. He attended their lectures and seems to have been most impressed by Heyne and Lichtenberg.[‡] Very early, Gauss became disenchanted with the mathematics professor W. Kästner, the well-known author of several then popular

---

* The decision in favor of Göttingen was not so exceptional. Ide, two years older than Gauss and also from Brunswick, preferred Göttingen, too. In both cases, the Duke continued his financial assistance.

† The fraternities (*Verbindungen*), later so famous and influential, arose only after 1815, after Napoleon had finally been defeated. They replaced certain older organizations which had existed at most of the German universities, but not at Göttingen. These older organizations were essentially regional in their makeup.

‡ Lichtenberg's name has survived in his many witty epigrams and aphorisms. He introduced the expressions "positive" and "negative" electric charge and the accompanying symbols + and − .

textbooks. Kästner, not a creative mathematician, was never accepted by Gauss as a teacher or colleague. He judged him with understandable but perhaps undue harshness, and even as an old man, long after Kästner's death, enjoyed ridiculing him.[3] In the later correspondence, Seyffer is depicted as an incompetent though socially pleasant colleague.*[4]

In any event, Gauss confirmed his decision to go to Göttingen by his frequent use of the library. The list of books he borrowed has been preserved; interestingly enough, it contains not only the expected mathematical literature but also contemporary novels, among them Richardson's *Clarissa*, which he read in English, and a Swedish grammar.[5]

Gauss did not make many friends among his fellow students. The only lasting relationship of which we know was formed with Wolfgang von Bolyai. Bolyai was a Hungarian nobleman from Transylvania, a province with a strong German minority. For us, the most important result of this association is the correspondence, extending over a period of more than fifty years, from 1779, during a temporary absence of Gauss from Göttingen, until 1853, two years before Gauss died. The letters are one of the main sources for our understanding of Gauss as a man and, mathematically, of the origins of non-Euclidean geometry. There is additional information in a short autobiography which Bolyai wrote for the Hungarian Academy of Science, and in a letter which he wrote to Sartorius, Gauss's first biographer, in 1855.[6]

Wolfgang von Bolyai, two years older than Gauss, enrolled in Göttingen in 1796 as a student of philosophy. This, at the time, included mathematics but Bolyai was genuinely interested in philosophical questions. His direct contact with Gauss lasted less than three years and ended when Bolyai returned to Hungary in 1799. His main mathematical interest was the foundations of geometry. In 1840, in an autobiographical sketch, he described how he became Gauss's friend:

. . . and I became acquainted with Gauss, who was then a student there [i.e., in Göttingen] and with whom I am still in friendly contact, though I could never compare myself with him. He was very modest and showed little; not for three days, as with Plato, but for years one could be with him without recognizing his greatness. It is a pity that I did not know how to open this silent, untitled book and read it. I did not know how much he knew, and he held me, after seeing what kind of person I was, in high regard without knowing how little I am. We shared the (invisible) passion for mathematics and our moral convictions; often we walked with one another, each of us occupied with his own thoughts, for hours not exchanging a word.[7]

---

* Gauss's judgments of Kästner and Seyffer seem to have grown harsher over the years. They are much less negative in the early correspondence with Bolyai. The turning point in his relation with Kästner seems to have been when the latter was not able to understand Gauss's theory of the division of the circle. Seyffer's incompetence became clear only later when Gauss worked as an astronomer. We will encounter similar shifts of opinion later.

The friendship was sealed by an exchange of tobacco pipes and the vow to smoke them daily at a certain hour and remember each other.[8]

Bolyai, though certainly not wealthy, belonged to another social class. He was a modest young man, full of enthusiasm for mathematics and admiration for his friend which he expressed openly and unconditionally. When they knew each other first, Gauss had not yet published anything, and his genius was far from obvious. Kästner, for example, appears to have been completely unimpressed by Gauss's abilities.[9] In June 1799 the friends saw each other for the last time (in this world) in the village of Clausthal, located halfway between Göttingen and Brunswick. Bolyai was on his way home and Gauss had suggested their meeting.

During his three years in Göttingen, Gauss studied entirely on his own terms. In the fall of 1798 he left the university, for reasons which are not clear to us; by then, he had already developed the basic ideas of nearly all his important mathematical papers, which he was to publish over the next twenty-five years. Gauss left Göttingen without a diploma. Following, as he wrote to Bolyai, a request from his Duke,[10] he submitted his doctoral dissertation to the University of Helmstedt in 1799. The degree was granted *in absentia*, without the usual oral examination.

In a letter to Bolyai, Gauss explicitly excluded Göttingen as the place for their last meeting, before Bolyai's return to Hungary. It probably would be wrong to suspect that Gauss had a hidden motive*—nothing was ever mentioned then or later, for instance in 1807 when Gauss returned as director of the observatory. It is more likely that when he went back to Brunswick Gauss was convinced that there was nothing left for him which he could learn in Göttingen, so he did not want to stay any longer.

Mathematically, the acquaintance with Kästner may not have been quite as inconsequential as it seemed. Kästner was an experienced teacher and a passable historian of mathematics. Probably influenced by, but certainly inferior to, Lambert, he was also interested in the foundations of geometry, i.e., in Euclid's system of axioms and the relation of the tenth (parallel) axiom to the others.[11] It may well be that Bolyai and even Gauss obtained some useful information and advice from Kästner.

---

* It is conceivable (and quite in line with similar reactions in comparable situations) that Gauss was afraid that this motive would be misunderstood (or, depending on one's viewpoint, be understood too well). This may have been what prompted Gauss to avoid Göttingen at the time.

The Organization of Gauss's Collected Works

The most visible fruit of Gauss's studies in Göttingen was the treatise *Disquisitiones Arithmeticae*. Published in 1801, it is Gauss's main number-theoretical work and one of the most important works in the history of mathematics. Before giving a summary of its contents we make a few methodological remarks.

The first source for this and the subsequent reviews of Gauss's other scientific work is the Academy Edition of Gauss's collected works (*Carl Friedrich Gauss' Werke*, occasionally to be cited as *G.W.*). Published in twelve volumes between 1863 and 1929, it is arranged according to a plan conceived soon after Gauss's death and completed under the direction of Felix Klein, who himself did not live to see the publication of the last volume. *G.W.* contains not only material actually published by Gauss, but also a reasonably complete* selection of his unpublished and unfinished papers, fragments and sketches never meant for publication, as well as excerpts from the correspondence. The first seven volumes are topical and cover the following: number theory (Vols. I and II), analysis (Vol. III), probability theory and geometry (Vol. IV), mathematical physics (Vol. V), and astronomy (Vols. VI and VII). Volume VIII contains various addenda. Volume IX is a continuation of Vol. VI and is devoted to geodesy. Volumes X and XI are actually in two parts, the first parts containing miscellaneous papers and documents and the second parts extensive and detailed essays by a number of very competent mathematicians and scientists who review Gauss's contributions to the fields in which he worked. Though of varying quality and interest, they are still the best guide to Gauss's work. Volume XII contains various short papers and a reprint of the atlas of terrestrial magnetism which Gauss compiled and edited together with Wilhelm Weber and his assistant W. Goldschmidt.

---

* Not literally, but certainly with regard to mathematical relevance. Only an antiquarian could be unhappy about the selection, but he might equally be satisfied that there is still some work left for him.

There are two reasons for the somewhat unsystematic organization of the collected works. Not all the papers in Gauss's estate were known of when the first volumes were prepared for publication; moreover, the original editor, Schering, died in 1897, i.e., between Vols. VI and VII, according to the chronology of *G.W.* The subsequent six volumes had several editors, most of them specialists in the areas to which the respective volumes were devoted.

Among the material which Gauss never meant to publish is the "mathematical diary",* a document that was accidentally found in 1898, more than 40 years after his death. It has entries from 1796 to 1814 but continuously covers only the years to 1801. With it, many of Gauss's results can be dated quite accurately; moreover, it provides some insight into his world of thought during his most productive years. Most of the 146 entries concern questions from analysis, algebra, and number theory; all are rather curt statements without proof or explanation. It appears that Gauss considered the diary as a sort of logbook for the documentation of his most meaningful or attractive discoveries. The first entry concerns the constructibility of the 17-gon; other important topics are the expansion of series into continued fractions, the decomposition of numbers into sums either of squares or of "triangular"[†] numbers, the law of quadratic reciprocity, the division of the circle and the lemniscate, the summation of certain integrals and series, the parallelogram of forces, the movements of comets and planets, the foundations of geometry, a formula for the determination of Easter up to the year 1999, and the arithmetico-geometric mean, a central concept in Gauss's investigation of elliptic integrals. It would be wrong to assume the diary actually reflected Gauss's development (it is too short and apodictic for that, obviously serving a different purpose); rather it shows what questions or what kind of questions Gauss was interested in, and how he valued his own results. As an historical source the diary is not reliable—the entries are not always clear, and some of the dates are more doubtful than they appear.[1]

Especially in his youth, Gauss had to be much more economical with paper than we would find convenient today. Some of his more important results occur on the margin and on interspersed blank pages of a book on elementary arithmetic (which we will cite, after its author, as *Leiste*). Gauss purchased his copy of *Leiste* well before going to Göttingen, and used it as a notebook. The entries in *Leiste* and in his regular notebooks (*Schedae*) are not reprinted in full in *G.W.*, but much of the information in these notes is of antiquarian interest only. The majority of Gauss's manuscripts are, of course, lost, but the remaining material suffices, in methodological terms, and allows us to reconstruct the major developments in reasonable detail.

---

* *G.W.* X, pp. 483–574

† To be defined and explained below, p. 33.

The collected works provide all the information which one needs to understand and appreciate Gauss's work. The archives[2] contain very little that is unpublished and that could contribute in an essential way. In Appendix A we discuss some minor details of the organization and authenticity of *G.W.* and of the editions of the correspondence. Although, as we shall see, some objections will have to be made, there is no reason to be particularly distrustful.

Most of Gauss's correspondence, and certainly everything important, is accessible in good editions. Only a small part of the letters is devoted to the discussion of mathematical questions, but what is, is of considerable importance. A series of letters to Bessel contains the statement, a proof, and a discussion of Cauchy's integral theorem[3]. This is probably the best-known mathematical passage in the correspondence, but there are many other instances of note.

Even if *Disqu. Arithm.* were not Gauss's first major publication there would still be good reasons to start an account of his mathematical work with number theory, and to begin an account of his number theoretical work with *Disqu. Arithm.* Gauss called number theory the "queen of mathematics"; for him, it was the first and most important part of mathematics, which he in turn called the "queen of science"[4]. Beyond this, there are methodological reasons for beginning here: arithmetic stands as an example for Gauss's other mathematical and scientific work. Even his work in applied mathematics and in fields like astronomy strives for the conciseness which accompanies and has to accompany number theory.

Chapter 3

The Number-Theoretical Work

Disquisitiones Arithmeticae was published in Leipzig, then the center of the German book trade. This was in the summer of 1801, nearly three years after Gauss had moved back to Brunswick. The following review will not be very ambitious, and we shall limit ourselves to a summary of Gauss's work, rather than try to evaluate it and its role in the development of number theory.

Disqu. Arithm. is divided into seven chapters, here cited as "sections" in accordance with the Latin original. Of these, the first three are introductory, Sections IV–VI form the central part of the work, and Section VII is a short monograph devoted to a separate but related subject. The work is dedicated to the Duke of Brunswick, without whose help it would never have appeared. The preface places the work in the tradition of number theory ranging back to antiquity; Diophantos, Fermat, Euler, Lagrange, and Legendre are named.*

The first section, only five pages long, deals with elementary concepts and results, such as the derivation of the divisibility rules for 3, 9, and 11. As the most basic concept in the work, congruences for rational integers modulo a natural number are defined and their elementary properties proved, among them the division algorithm.

In Section II (24 pages), Gauss proves the uniqueness of the factorization of integers into primes and defines the concepts "greatest common divisor" and "least common multiple". After defining the expression $a \equiv b \bmod c$, Gauss turns to the "equation" $ax + t \equiv c$.† He derives an algorithm for its solution and mentions the possibility of using continued fractions instead of the Euclidean algorithm. Another topic is Euler's totient function $\varphi(m)$; it denotes the number of the integers that are less than m and prime to m.

---

* Much later, in a letter to the astronomer Schumacher, Gauss expressed his low opinion of Diophantos more explicitly than in the cryptic allusion of the preface.

† Gauss introduced this symbol; the concept, of course, is much older.

$\varphi(m)$ is studied with what amounts to the multiplicative properties of the primitive residues.

The title of Section III (35 pages) is "De Residuis Potestatum"; it contains an investigation of the residues of a power of a given number modulo (odd) primes. The basis of the investigation is Fermat's famous "little" theorem

$$a^{p-1} \equiv 1 \pmod{p}, \qquad p \text{ a prime which does not divide } a.$$

Gauss gives two proofs, one by "exhaustion" which goes back to Euler or possibly even to Leibniz, the other, also not essentially new, uses the "binomial theorem"

$$(a + b + c + \cdots)^p \equiv a^p + b^p + c^p + \cdots \pmod{p}.$$

This leads to the notion of primitive root: a is a primitive root, if the powers a, a^2, a^3, $\cdots$ yield (mod p) all intergers relatively prime to p; using this concept Gauss defines the index e of a number b rel. to a by

$$a^e \equiv b \pmod{p}.$$

In this relation, a is a fixed but arbitrary primitive root (*Basis*) mod p. Gauss shows how to make computations with the help of these indices which he compares to logarithms.[1] For the benefit of the reader, a table of indices which he had compiled is added in an appendix. From the index representation Gauss obtains a criterion for the quadratic character of a number (i.e., whether a number is or is not a quadratic residue mod p). This theorem was already known to Euler, but Gauss's derivation and proof are more complete and conclusive. Another consequence is Wilson's theorem

$$1.2.3 \ldots (p-1) \equiv -1 \pmod{p}. \quad [2]$$

Essentially, Sections I–III of *Disqu. Arithm.* constitute a systematic introduction to elementary number theory and prepare the reader for the main part of the book, Sections IV and V.

The central topic of the fourth section (47 pages) is the law of quadratic reciprocity. The law derives its name from a formalism invented by Legendre and defined as follows: Let p,q be positive, odd primes. Then $(\frac{q}{p})$ is $+1$ if $x^2 \equiv q \pmod{p}$ is solvable in whole numbers, and -1 otherwise. The law of quadratic reciprocity is the identity.

$$\left(\frac{p}{q}\right)\left(\frac{q}{p}\right) \equiv (-1)^{\frac{p-1}{2} \cdot \frac{q-1}{2}} \qquad [3]$$

This is not a formalism used by Gauss (nor did he use the expression "quadratic reciprocity") though clearly this is the best way to express the relation between $(\frac{p}{q})$ and $(\frac{q}{p})$. The theorem itself had been formulated by Euler and discussed at length by Legendre but it had not been proved correctly. Gauss's proof starts with heuristic considerations in which he shows that the law holds for selected primes. After these inductive beginnings, Gauss proceeds to treat the general case by complete induction over the primes. This first of Gauss's proofs (he gives essentially six different ones) is quite labored and contains the separate treatment of eight different cases, but uses only elementary tools.*⁴ Using Legendre's formalism, Dirichlet simplified this proof, reducing the total number of cases to two. Gauss, calling it the *theorema fundamentale*, expressed the law the following way (§131): If p is a prime of the form $4n + 1$, then $+ p$ will (will not) be the residue of any prime which, taken positively, is the residue (not the residue) of p. For prime numbers of the form $4n + 3$, an analogous statement holds with $-p$.†⁵

Directly after proving the fundamental theorem, Gauss solved, by factorization, the problem of determining for two arbitrary numbers P,Q whether Q is quadratic residue modulo P. The most general formulation, in this context, of the law of quadratic reciprocity is contained in §146. Another application, subsequently given by Gauss, is the construction of linear forms containing all primes which are or are not quadratic residues of a given number. Section IV ends with the reduction of nonpure congruences of the second degree, i.e., congruences of the form $ax^2 + by + c \equiv 0 \pmod{p}$, to pure ones. In a short historical review of Euler's and Legendre's work Gauss points out that his is the first correct and complete proof, but gives credit, without going into details, to the work of his predecessors.⁶

The fifth section (260 pages) is the central part of *Disqu. Arithm.* It deals with the theory of binary quadratic forms, i.e., algebraic expressions of the type

$$f(x, y) = ax^2 + 2bxy + cy^2 \qquad (a, b, c \text{ given integers}).⁷$$

A substantial portion of the fifth section is not original but repeats and summarizes results which are due to Lagrange. Gauss indicates where his original work begins, and we shall make a corresponding remark in the course of our summary. Gauss's algebraization of arithmetic leads to quite complicated algebraic computations and concepts, without direct number-theoretical motivation. Later, we shall see how Gauss reestablishes this connection whenever necessary and possible. For him, arithmetical questions are at the center of the investigation, and not their abstract algebraic theory.

---

* Gauss found the proof in the spring of 1796, after lengthy efforts. At this point, Gauss was not yet able to see it within the framework of a more general theory, and his proof was the result of an incredibly energetic effort which would have to be followed step by step to be understood (and appreciated). The critical point was the proof of the fact (which is necessary for the proof of the theorem) that primes $p > 5$ and of the form $4n + 1$ are always nonresidues of a smaller prime. This is easy for $p \equiv 5 \pmod 8$, but very difficult for $p \equiv 1 \pmod 8$. This last case cost Gauss a whole year.

† Dirichlet formulated the theorem in the following form: Let p,q be two positive, odd prime numbers of which at least one is of the form $4n + 1$. q is quadratic residue or nonresidue of p whenever p is a quadratic residue or nonresidue of q. If p and q are both of the form $4n + 3$, then q is quadratic residue or nonresidue of p whenever p is a quadratic nonresidue or quadratic residue of q.

In the first paragraphs, Gauss discusses two basic algebraic properties of a quadratic form, the identity

$$f(x, y)f(x', y') = [(ax + by)x' + (bx + cy)y']^2 - D(xy' - x'y)^2$$

with

$$D = b^2 - ac$$

and the relation

$$D' = D(\alpha\delta - \beta\gamma)^2$$

where D' is the discriminant[8] of a form $F' = a'x'^2 + 2b'x'y' + c'y'^2$. One obtains F' from F by a linear transformation of the variables with coefficients $\alpha,\beta,\gamma,\delta$; D is the discriminant of the form F. The two major problems are to find all possible ways of representing a given number by a given form and to review all the representations which belong to one value or to different values of a form. A transformation law for forms follows. If a form F can be transformed into a form F' by a substitution with integer coefficients, F' is said to *contain* F. Then the discriminant of F' divides the discriminant of F. F and F' are *equivalent* if F' contains F and F contains F'; they have the same discriminant. This leads to a classification of the forms induced by the determinants of the transformation matrices. Moreover, one distinguishes between *properly* and *improperly equivalent* forms; forms are improperly equivalent if the determinant of the transformation matrix has the value -1. The classification of forms is one of the most useful tools in the subsequent investigation.

Properties which do not depend on the type of form are its first object, among them criteria for the equivalence of forms and, as a concrete application, the representation of numbers by forms. Of special interest are the *ambiguous* forms, i.e., forms which are improperly equivalent to themselves. The existence of ambiguous forms is proved, and Gauss also derives some of their elementary properties which are needed later. Forms with positive square discriminants are trivial, so Gauss proceeds to forms with negative discriminant. He shows that each class of properly equivalent forms with a given discriminant can be represented by a certain well-defined normal form, the *reduced form*. This is a first step of the subsequent exhaustive treatment of the algebraic characteristics of such forms. Gauss provides criteria for the equivalence of two forms, calculates a bound for the number of different reduced forms with given negative discriminant, derives the transformation law for properly equivalent forms, and solves the task of finding all transformations between them. The necessary computations are extensive but can be shortened with the help of a table of reduced forms. Finally, Gauss proves several decomposition theorems for prime numbers, among them the unique decomposition of primes of the form $4n + 1$ into sums of two squares (§182).

Gauss now turns to forms with given positive nonsquare discriminants. He again succeeds in defining a normal form which, this time, is not unique. The set of reduced forms (i.e., normal forms) that belong to a given form is finite and has certain algebraic properties; it is called the *period* of reduced forms. Gauss shows that equivalent forms with a given discriminant are uniquely characterized by their period. Furthermore, these periods again induce a classification of the forms. In his exposition Gauss proceeds constructively and provides the reader not only with detailed instructions of how to find the reduced forms, but also with several numerical examples. His considerations are concluded by the theorems which describe the transformation of equivalent forms into each other. This leads to *Pell's equation*

$$t^2 - Du^2 = 1$$

and to a generalization of it (§201). Pell's equation had already been investigated by Fermat and his contemporaries; its solutions provide all required transformations between equivalent forms if only one transformation is known.

Having advanced so far, Gauss points out that all the results hitherto proved in Section V had rigorously been proved by Lagrange; in general, this first part of the fifth section follows the tradition of Fermat and Lagrange. The theory is rounded off by a short treatment of forms with positive square and with zero discriminants where analogous results can be obtained immediately. To end, Gauss solves the general diophantine equation mod p of the second degree in two unknowns. Gauss finally determines the distribution of all forms with a given discriminant over a finite number of classes of equivalent forms. These investigations of the theory of forms give a first insight into the way a given number can be represented. This is the initial topic of Section V; the algebraic considerations are directly used for arithmetical purposes.

The paragraphs which follow are devoted to a closer investigation of the algebraic properties of forms; the concepts *order, genus,* and *character* are introduced and discussed.* Gauss's basic problem is the characterization of the numbers which can be represented by classes belonging to a given discriminant D.

The results which now follow are original and had a lasting influence on the development of number theory. The order of a class is determined by the greatest common divisor of the coefficients of the forms in a class. A *primitive* order consists of forms whose coefficients are mutually prime, a *properly primitive* order of forms is one where $(a, 2b, c)$ are pairwise relatively prime. If a class contains one (properly) primitive form all its forms are (properly) primitive; one can consequently speak of *(properly) primitive classes*. Two classes are in the same *order* if the coefficients of any two representative forms have the same greatest common divisor. All the primitive classes form a certain order, as do the properly primitive classes.

---

* Later, in the literature after Gauss, ideal classes in quadratic number fields replace the much less convenient forms. The transition from forms to classes is not quite straightforward, because some information which is provided only by the forms is lost in the process.[9]

Orders can be classified by genera; this is based on the following (§229): Let F be a primitive form with discriminant D and the prime p a factor of D. All numbers which can be represented by F and which do not contain p are either quadratic residues or nonresidues of p. This determines the form F completely. The set of all the special characters of a form or of a class is called the *total character* of this form or class. All classes with the same total character are considered to be in the same *genus*. In an example, Gauss shows that one obtains for the discriminant -161 certain properly primitive classes which can be assigned to four different genera. The form $(1, 0, -D)$ is the principal form, its class the principal class, and its genus the principal genus.

With §234, the famous subsection "On the Composition of Forms" starts. In it, and in subsequent subsections, Gauss treats the composition of classes, orders, and genera. He then develops a theory for these concepts and shows how the genera of two forms determine the genus of the composite form, etc. "On the Composition of Forms" can be considered the centerpiece of *Disqu. Arithm.*; it quickly attained a reputation of depth and inaccessibility. Expressed in modern terminology, the properly primitive classes form an abelian group, with the "principal class" represented by the form $x^2 - Dy^2$ as the unit element. Although the computations are difficult and interesting, Gauss does not concern himself with any abstract algebraic relations. He immediately proceeds to the arithmetical substance of the theory. Among the most important results are theorems about the class numbers in genera of the same order and of different orders and about the number of ambiguous classes for a given discriminant. The crowning achievement of these concepts is a new proof of the law of quadratic reciprocity, Gauss's *theorema fundamentale* (§262). It is based on the fact that the prime genus will correspond to one of the characters if only two characters exist for a given non-square discriminant. Of these two characters one will not correspond to any properly primitive form of the discriminant. This is true because half of the characters belonging to a discriminant do not have properly primitive genera, which is proved in §261. Gauss concludes this subject with another concrete application, obtaining a way of decomposing primes into sums of two squares. This is possible because all ambiguous properly primitive classes of a given discriminant p are equivalent if p is a prime number of the form $4n + 1$.

At this point, Gauss begins an excursion into the theory of ternary forms. This is needed for the calculation of the exact number of genera for a given form.[10] A *ternary form* is an expression

$$f = ax^2 + a'x'^2 + a''x''^2 + 2bx'x'' + 2b'x''x + 2b''xx'$$

(a, b integral numbers)

with the discriminant

$$\Delta = ab^2 + a'b'^2 + a^2b''^2 - aa'a'' - 2bb'b''.$$

Gauss starts with a derivation of elementary transformation properties, defines equivalent forms, and studies equivalence classes of ternary forms. He recognizes as the four principal problems in the theory of ternary forms (i) to find all representations of a given number by a given form, (ii) to find all representations of a given binary form by a given ternary form, (iii) to find a test for the equivalence of two ternary forms and then to find the transformations between two equivalent forms, and (iv) to decide whether a given ternary form contains another ternary form with a greater discriminant and, if it does, the corresponding transformation. Gauss's program is to reduce (i) to (ii), (ii) to (iii), and to solve (iii) for some important cases. Problem (iv) is listed but not treated. Gauss confines himself to special cases and avoids full generality as his computations appear to have been too complicated to generalize further. Gauss's later work contains more about ternary forms and we shall come back to them below.

Utilizing his newly obtained results, Gauss represents binary by ternary forms. Such a representation is always possible by using the integral substitution

$$x_i = \alpha_i t + \beta_i u, \qquad i = 1, 2, 3.$$

By considering the representation of a binary form by the special ternary form $x_1^2 - 2x^2 x^3$, the existence of genera for exactly half of the total characters is proved. It is possible to calculate the characters with the help of the law of quadratic reciprocity.

Among the applications of this part of the theory are the representation of numbers (and of binary forms) as sums of three squares (§291) and a proof of Fermat's statement—which had until then not been proved—about the decomposability of an arbitrary positive number into a sum of three triangular numbers (see p. 33). A third consequence is another of Fermat's results, namely that every integer can be represented as the sum of at most four squares.

The introduction of ternary forms not only provides a tool for the solution of certain problems in the theory of binary forms, it also marks the beginning of a rich new field of mathematical investigation, which is associated with the names Dirichlet, Eisenstein, H. J. S. Smith, and Minkowski.

At this point, Gauss analyzes Legendre's (unsuccessful) proof of the *theorema fundamentale* and shows why it is not complete.

Gauss proceeds (without proof) to propositions about the mean number of genera and classes for a given discriminant D. His asymptotic formula for the first is

$$\alpha \log |D| + \beta, \qquad \alpha, \beta \text{ constant}$$

and for the second

$$\gamma \sqrt{|D|} - \delta, \qquad \gamma, \delta \text{ constant.}$$

The latter formula holds only for negative D; an analogous formula holds for positive, nonsquare D, but Gauss was not able to determine the constants in this case. His results appear to have been based on comprehensive numerical computations, although he drops some hints about deep theoretical considerations which were to be explained later.

The fifth section ends with remarks about the classes which belong to the principal genus of forms with a given discriminant. Their cyclic character leads to the definition of periods of the class C if C is an arbitrary fixed class in the principal genus. Discriminants are called *regular* if their principal genus is contained in a single period (i.e., it can be represented by a finite sequence $C, C^2, C^3, \ldots, C^{n+1} = C$), otherwise they are irregular. Gauss does not pursue the theory here and only adds a few remarks about the relation between the discriminant and the corresponding degree of irregularity or the corresponding regular case. The very last remarks of the section are devoted to an explanation of a method to find all properly primitive classes for a given regular discriminant; in an example, Gauss calculates the genera and classes for the discriminants -161 and -546.

The sixth section of *Disqu. Arithm.* will be discussed only briefly. In it, Gauss presents several important applications of the concepts of Section V which were not included in that section. The principal topics are partial fractions (i.e., the decomposition of a fraction into a sum of fractions with the prime factors of the original denominator as denominators of the summands), periodic decimals, and the resolution of congruences by Gauss's "exclusion" method. Another interesting topic is the derivation of criteria for distinguishing between composite numbers and primes. Section V, with the sixth section as an appendix, constitutes a natural conclusion to *Disqu. Arithm.*

Dirichlet's lectures, edited and augmented by Dedekind, are virtually a running commentary of the first four or five sections.[11] They are still of interest as a (slightly ahistoric) introduction to Gauss's work. A useful complement to Dirichlet-Dedekind is the book on quadratic fields (*Quadratische Körper*) in the third volume of Weber's *Algebra* (1899).

Section VII is the most popular part of *Disqu. Arithm.* Its historical influence was enormous—we only mention §335 with its hint, so important to Abel, about the possible generalizations of the division of the circle.* Interchapter IV contains an extensive summary, to illustrate Gauss's mathematical style. There are specific reasons why the seventh section is particularly suitable for such a purpose. It is homogeneous and basically self-contained; most of it was written before Section V had been completed.

The division of the circle with ruler and compass is an old and classical problem. Since antiquity, it had been an open, often discussed question whether the regular 17-gon could be constructed with the help of these two

---

* The next step is the division of the lemniscate, a problem which Gauss had solved before the completion of *Disqu. Arithm.* as we know from his diary and other substantial concurring evidence.

tools only. In his theory of the division of the circle, Gauss solved this specific case and the general constructibility problem for regular n-gons, n a prime number > 2.

Cyclotomy, the theory of the division of the circle, is concerned with the equation

$$x^p - 1 = 0, \qquad p \text{ an odd prime.} \qquad (*)$$

The roots of $(*)$ lead to trigonometric functions of the angles $2\pi k/p$ $k = 0, 1, 2, 3, \ldots, p - 1$. This was well known at the time and had already been discussed by Euler.

The two basic ingredients of Gauss's proof are the use of primitive roots and a skillful manipulation of the way the coefficients of a polynomial can be expressed as (symmetric) functions of its roots. Primitive roots can be used because the powers of an arbitrary primitive root mod n correspond to the $p - 1$ roots of

$$X \equiv \frac{x^p - 1}{x - 1} = 0.$$

Gauss's presentation is careful and contains many detailed numerical examples. The final result of the section is in §365 where Gauss shows that any regular n-gon with n a prime number of the form

$$2^{2^\nu} + 1 \qquad (**)$$

can be constructed with straightedge and compass. Only if n has the form $(**)$ can the polynomial X be reduced to a sequence of quadratic equations, which in turn is what permits geometric solution.

Dedekind's booklet [1901] gives a detailed and readable account of the contents of the seventh section. Similar to the other sections of *Disqu. Arithm.*, the seventh section contains several additional arithmetical results which follow from the general theory. In §356, there is a paragraph about what are now called Gauss sums, expressions which appear here for the first time in Gauss's work. In *Disqu. Arithm.*, Gauss made only a few remarks about them but devoted a later paper to a deeper investigation, "Summatio quarundam serierum singularium" (1808). It will be summarized below. *Disqu. Arithm.* ends with several tables which Gauss compiled for the benefit of the reader.

Gauss repeatedly mentioned in his correspondence and in other publications that he intended to write a sequel to *Disqu. Arithm.* It is easy to see why this never happened; the circumstances of Gauss's life provide ample explanation. Though we do not exactly know what Gauss intended to include in such a second volume, we have a fairly good idea from the shorter number-theoretical papers which Gauss published and from the manscript *Analysis Residuorum*, a fragment which became known only after Gauss's death. *A.R.* is the draft of a major work and contains an earlier version of the last part of *Disqu. Arithm.*, together with some additional passages which we specifically refer to here. Not all of *A.R.* was included in *G.W.*; the editors deleted whatever was substantially contained in *Disqu. Arithm.*

Instead of completing *A.R.* or the projected other major arithmetical work, Gauss confined himself to a sequence of shorter papers, the last of which appeared in 1831. This paper is the second part of an essentially fragmentary work on the theory of biquadratic residues, continuing an earlier paper of 1825. The subjects of the other papers are Gauss sums (1808) and further proofs of the law of quadratic reciprocity. In our review, we can confine ourselves to three main topics: (i) further proofs of the law of quadratic reciprocity, (ii) Gauss sums, and (iii) the theory of cubic and biquadratic residues. We will treat (i) last because a discussion of the different proofs of the law of quadratic reciprocity will naturally lead us to a summary of Gauss's arithmetical work.

The "singular series" of the title of "Summatio quarundam serierum singularium" are actually Gauss sums, i.e., expressions of the form

$$W = \sum_{v=0}^{n-1} e^{\frac{2\pi i}{n}v^2}.$$

Though only playing a marginal role in Gauss's work, these sums became much more important in the subsequent development of number theory. Gauss was led to the contemplation of ϑ-series and to a variety of interesting results and relations by them. He did not publish any of this work, which will be discussed in two places here, directly below and on pp. 87ff.[12] The object of *Summ. Ser.* is to determine the signs of the expressions (*) and (**), a problem which Gauss had not solved when he introduced them in *Disqu. Arithm.* In the following discussion, one has to distinguish two cases, depending on p:

(i) For p of the form $4m + 1$,

$$\sum \cos ak\omega = -\tfrac{1}{2} \pm \tfrac{1}{2}\sqrt{p},$$
$$\sum \cos bk\omega = -\tfrac{1}{2} \mp \tfrac{1}{2}\sqrt{p}, \qquad\qquad (*)$$

and consequently

$$\sum \cos ak\omega = \sum \cos bk\omega = \pm\sqrt{p},$$
$$\sum \sin ak\omega = 0,$$
$$\sum \sin bk\omega = 0.$$

(ii) For p of the form $4m + 3$,

$$\sum \cos ak\omega = -\tfrac{1}{2},$$
$$\sum \cos bk\omega = -\tfrac{1}{2},$$
$$\sum \sin ak\omega = \pm\sqrt{p},$$
$$\sum \sin bk\omega = \mp\sqrt{p}, \qquad\qquad (**)$$
$$\sum \sin ak\omega - \sum \sin bk\omega = \pm\sqrt{p}.$$

Gauss's tools are all elementary; the procedure is complicated though never difficult. There are many numerical examples, motivating explanations, and summaries. The last paragraphs contain a new proof of the law of quadratic reciprocity and several theorems which are termed "supplements to the law of quadratic reciprocity". These theorems determine the primes having residues $-1, 2, -2$.

We know from the correspondence how difficult it was for Gauss to determine the signs of the sums; in September of 1805, he wrote to Olbers:

> ... What I wrote there [Disqu. Arithm. §365] ..., I proved rigorously, but I was always annoyed by what was missing, namely, the determination of the sign of the root. This gap spoiled whatever else I found, and hardly a week may have gone by in the last four years without one or more unsuccessful attempts to unravel this knot—just recently it again occupied me much. But all the brooding, the searching, was to no avail, and I had sadly to lay down my pen again. A few days ago, I finally succeeded—not by my efforts, but by the grace of God, I should say. The mystery was solved the way lightning strikes; I myself could not find the connection between what I knew previously, what I investigated last, and the way it was finally solved.[13]

Gauss's results on biquadratic residues appeared in the papers "Theoria residuorum biquadraticorum I & II", both published in the journal of the Göttingen Royal Society. They are in Latin; more accessible are Gauss's German summaries. The first of the two papers contains minor, isolated results and concepts which are developed by analogy to the quadratic case. This prepares the reader for the central topic of the two papers, the determination of the biquadratic character of a given prime. For a general and efficient treatment, one has to use complex numbers, which Gauss introduces at the beginning of the second paper.

The two papers do not contain a general proof of the law of biquadratic reciprocity. Such a proof was first given by Eisenstein, long after Gauss had stopped publishing his number-theoretical results. In the two published parts of Th. res., Gauss formulated and proved the law for several important special cases, the numbers $\pm 1 \pm i$. The general theorem was to be the subject of a third part which was never written. Posthumously published material seems to indicate that Gauss actually possessed a proof (G.W. X,1, p. 65) but the fragment which contains it cannot be dated. So it is not clear whether this proof was original or based on Eisenstein's proof, to which it is very similar. It proceeds in analogy to the quadratic case which Gauss had developed within the framework of the division of the circle.

There is nothing systematic on the theory of cubic residues from Gauss's hand, though we know of efforts to prove the law of cubic reciprocity. There are some scattered results for certain primes, of methodologic and personal but not of arithmetic interest.

Of Gauss's unfinished and sketchy fragments, those are particularly interesting that contain results usually proved by analytic methods. Gauss knew, probably by induction, several asymptotic laws for the class number of forms and the distribution of quadratic residues. There is also a remark, dating from 1796, about the number of primes less than a according to which $a/\log a$ is an asymptotic bound (see also Interchapter III below).

Our review of Gauss's number-theoretical work has been ahistoric and does not try to trace the genesis of his various results and publications. It is easy to date the majority of Gauss's discoveries, even those that became known only after his death. Often the publication of his papers was considerably delayed but this does not pose any major problems either.*

We know of *Disqu. Arithm.* that the first four sections had been drafted by 1796 and had been written up in virtually final form by the end of 1797, i.e., during Gauss's second year at Göttingen. A first draft of the fifth section was completed in the summer of 1796. It went through several revisions until it assumed its final form in the winter of 1798/99, when the passages on ternary forms were added. It was probably completed during the first half of the year 1800. Sections VI and VII were less of a problem and did not have to be revised substantially.

Gauss's published papers give a good idea of his progress and development. Other sources are the many fragments, known only long after his death, the correspondence,[†] and the diary. Of course, many of his notes and drafts of his manuscripts are lost, but we have no reason to bewail losses irreplaceable in terms of content.

Though clearly of much value, even a document like the diary leaves many questions unanswered. The entries are very brief, and it is occasionally difficult, if not impossible, to understand them correctly. As an example, we quote the entry of April 8, 1796: "Numerorum primorum non omnes numeros infra ipsos residua quadratica esse posse demonstratione munitum".[‡] This is not a "deep" statement. It merely says that not all the numbers which are smaller than a given prime are quadratic residues modulo that prime. In their notes that accompany the diary, Klein and Bachmann point out that Gauss must have known this much earlier. From external evidence—a handwritten remark by Gauss in reference to the paragraph of *Disqu. Arithm.* in which the theorem was proved—they infer that Gauss completed that very day, April 8, 1796, his first proof of the law of quadratic reciprocity. Schlesinger, in his essay, goes a step further and quotes the entry as *evidence* for the fact that Gauss had actually proved the law on that day. The diary leaves us similarly confused with regard to Gauss's work about cubic and biqua-

---

* One can basically rely on the collected works and on Bachmann's comments in his essay in Vol. X,2. See also Appendix B.

[†] Which is much less informative at this stage of Gauss's career than later.

[‡] The Latin of this quotation is certainly not of classical quality—Gauss seems to have preferred a conversational silver to the classical gold.

dratic reciprocity. There are several entries from 1807 (##130–133) according to which he found his main results, not necessarily the proofs, during that winter. This agrees with a statement in a letter to S. Germain of the same year (April 30, 1807) but not with the remark in the published paper on biquadratic residues (Part I), according to which he already had his results in 1805 and even less with entry #144 of October 23, 1813, the day on which his youngest son Wilhelm was born. In this entry, Gauss explicitly stated that he had found, after seven years of fruitless efforts, the general basis for the theory of biquadratic residues. This sheds new light on his earlier optimistic and, more importantly, public assertions, which seem to indicate that Gauss considered much of the work done once he had succeeded in the explicit formulation of the results (or, better, the scope) of his investigation.

A more positive example is entry #18 of July 10, 1796. It concerns the decomposability of every number into the sum of three triangular numbers:*

$$\text{EYPHKA! num} = \triangle + \triangle + \triangle$$

Obviously, the diary is only of limited usefulness for a history of Gauss's development. This should not come as a surprise—the picture of Gauss faithfully entering each night the day's discoveries is odd and improbable. Still, the diary is a stimulating and revealing document. Its informal statements are like markers in a waterway and permit us to guess how Gauss's thinking developed. But his most formative years are hidden from our view, the long years of reception, of unending calculations and aimless manipulations of numbers, of the computation of curious tables, and of unsystematic heuristic efforts. Only sudden discoveries, recurring examples, and favorite topics provide rare glimpses of insight, but there is no key that would allow us to decipher Gauss's thought systematically and read his mind easily.

We now return to the law of quadratic reciprocity. Altogether, Gauss proved the law in six different ways; he himself, as we shall see later, distinguished eight proofs. The two proofs in *Disqu. Arithm.*, though quite different, both date from 1796 (April and June). The first of them, the most elementary among Gauss's proofs, is lengthy and complicated, but direct, and uses only elementary methods. The second proof, formally much easier, is part of the theory of quadratic forms, which is the subject of the fifth section of *Disqu. Arithm.* Chronologically, an (by Gauss) unpublished proof comes next, the one in *Analysis Residuorum*. Gauss actually speaks of two proofs, but they are quite similar and follow from his *theorema aureum*. This theorem, which he himself called "golden", states the existence of an equation, modulo any prime p with $p - 1 = e.f$, such that the roots of this equation are the e periods, each consisting of f members, of the pth unit roots. Gauss's proof in *A.R.* also dates from 1796, but it is possible that he had proved the *theorema aureum*

---

* Triangular numbers are numbers of the form $\frac{1}{2}s(s + 1)$. The proposition had already been stated by Fermat. Gauss proved it in §293 of *Disqu. Arithm.*.

even earlier. There is a corresponding entry on the margin of *Leiste*, formulating but not proving it, which seems to be early but cannot be dated reliably. Also, entry #39 (of October 1, 1796) of the diary refers to it—another instance where seemingly precise information from the diary does not help.

The proof in *Summ. ser.* is from 1801 but was published only much later, in 1808. It is an important proof because using the theory of Gauss sums one can easily calculate the number of quadratic residues or nonresidues in the sequence $1, \ldots, p$. The fourth proof, in the paper "Theorematis fundamentalis in doctrina . . . " (published in 1817), is another offspring of cyclotomy, starting out with the basic expression

$$x - x^g - x^{g^2} - \cdots - x^{g^{p-2}},$$

where g is a primitive root of the given prime p ($\neq 2$). The proof itself is not difficult, but it makes essential use of the theory of higher congruences. The other two proofs were found between 1805 and 1810, probably in 1807 and 1808. They are part of the theory of cubic and biquadratic residues. Along with the second of these proofs Gauss develops an algorithm for the determination of the quadratic character of a given number with respect to another number. So we see that all of Gauss's proofs were probably found within a period of twe ve years, between 1796 and 1808; those that were published during Gauss's lifetime were published between 1801 and 1818. Some of these proofs are quite late and come in a period when Gauss was deeply involved in astronomical research. This, but even more clearly the fact that he continued to work in number theory and to publish well beyond 1815, shows that Gauss did not stop concerning himself with pure mathematics after he became engrossed in astronomy and other applications.

It may in retrospect be possible to discern the decisive ideas in a theory and to isolate them from the routine details accompanying the completion of a proof or a theory, but the creative involvement does not stop for a mathematician—and did not for Gauss—before everything has been worked out completely and a string of details have been developed into a theory. Involved as he was in his astronomical efforts Gauss did not force himself to drop his original interests completely, though there were occasional complaints in the correspondence that he did not have enough time and leisure to pursue them.[14] Gauss may have enjoyed occupying himself simultaneously with several diverse subjects—the correspondence and the abrupt changes of topic in the diary make this quite likely. Throughout his scientific life, well into the 1840s, Gauss followed the literature in number theory closely and was well aware of the contributions of his younger contemporaries Jacobi, Dirichlet, and Eisenstein.

It seems appropriate to discuss in this context the origins and nature of Gauss's historical remarks. They occur in almost all his mathematical works, but are most frequent and relatively systematic in *Disqu. Arithm.* It is quite obvious that these remarks are neither reliable nor complete, even with respect to the literature which Gauss knew well. Gauss had a certain "natural" historical interest, but this was predominantly mathematical and personal. Even early in his life he saw himself as a historical figure whose work would be studied by succeeding generations of mathematicians. This is an attitude which is reflected in Gauss's historical remarks, and it would be futile to scrutinize them with the instincts of a historian or antiquarian. Even priority conflicts were irrelevant to Gauss, as long as they were purely formal, without discussions of mathematical substance. Gauss saw himself as a member of the sequence of the great number theorists which started with Diophantos and which included Fermat, Euler, and Lagrange; Gauss's concern for their work seems to have been limited by the degree to which he could absorb it into his own world of ideas and eventually render it obsolete. This explains his principal disinterest in any complete historical documentation and the ensuing incompleteness of his references. Gauss seems to have been free of any historical bias, and his references were certainly intended to give a fair survey of the relevant work of his predecessors and contemporaries. He was in fact quite successful at this, for his historical work is better than its reputation. Examples are the reference to Euler as the discoverer of the law of quadratic reciprocity, a fact which is occasionally overlooked even today, and several remarks that give more credit to Legendre than one would expect after the very critical comments in *Disqu. Arithm.*[15]

None of Gauss's results in number theory was published before 1801, with the exception of the announcement of the constructibility of the 17-gon. Though we do not aim to give a "linear" presentation of Gauss's life we so far have not deviated from this (questionable) ideal. The period in which number theory was the focal point of his scientific interests ended soon after his return to Brunswick, and arithmetic was replaced by astronomy. As in Gauss's life, number theory will be a recurring topic in this biography; specifically, Section VII of *Disqu. Arithm.* will be used in an attempt to analyze Gauss's mathematical style and methodology (see Interchapter VI). This will be our second attempt—the preceding summary should already have given some idea not only of Gauss's results and discoveries but also of the way in which he developed and explained a major area of mathematics.

The Influence of Gauss's Arithmetical Work

Disquisitiones Arithmeticae and Gauss's other papers in number theory, including much that was published posthumously, had an enormous and lasting impact on the development of the theory of numbers in the 19th and first half of the 20th century. It is said of Dirichlet that he always had a copy *Disqu. Arithm.* on his desk; he studied it religiously, not an altogether unreasonable way to learn arithmetic. Why Gauss's ideas played such a central role is easy to see. Gauss's dominant interest in concrete problems and his reluctance to employ abstract concepts led him to create the ideal tools for his "theoretical" summary of older results and to discover such a rich variety of new concepts.

Combinatorics has so far never been mentioned as an area that Gauss was interested in. Though fond of performing lengthy calculations and of numerical work in general, combinatorics in the sense we understand the word today seems never to have been very attractive to Gauss; for him, numbers and their properties were intricately (and probably exclusively) connected with their arithmetical theory. To Gauss, "theory" appeared to be much less abstract than it might have been to other, less numerically oriented mathematicians. This explains why one encounters so many special cases of concepts and notions in Gauss's work that reemerge in the work of the mathematicians of subsequent generations.

We proceed to give several examples of the way Gauss's work contains the germs of later theories. They are of intrinsic interest, but also shed new light on Gauss's number-theoretical work.

We start with an example which readily comes to mind: the several asymptotic laws which Gauss seems to have found through a combination of heuristic and theoretical considerations (cf. p. 27). Among his manuscripts, there is the following posthumously published list of propositions:

Prime numbers less than a

$$\frac{a}{\log a}$$

numbers with two factors

$$\frac{\log(\log a)a}{\log a}$$

with three factors

$$\frac{\frac{1}{2}a(\log \log a)^2}{\log a}$$

$$\vdots$$

The general formula was not proved until 100 years after Gauss conjectured it, by Landau in *Bull. Soc. Math.* **28**, 1900. A discussion of this and of Gauss's other asympototic laws can be found in Vol. X of *G.W.*, pp. 11–18.

That Gauss had far-reaching numerical conjectures may not be as surprising as the many theoretical concepts which he preempted in his work and of which we shall now consider a few examples.

Disqu. Arithm. § 272 contains the reduction theory of ternary forms and culminates in a theorem which gives estimates of the coefficients of a "minimal" equivalent form in terms of the discriminant of the original form. This led Hermite (also Korkiné, Zolotareff, and others) to a reduction theory of *n*-ary positive quadratic forms after Hurwitz had proved that it suffices to consider the reduction theory of positive forms. Hermite's proof was analogous to the one by Gauss for ternary forms; later, this work was generalized by Minkowski in his famous book on the geometry of numbers, where he gives the following estimate:

$$\frac{M}{\sqrt[n]{D}} < 4 \, \frac{\Gamma\left(1 + \frac{n}{2}\right)^{2/n}}{\Gamma\left(\frac{1}{2}\right)^2}.$$

M is the minimum of the form under consideration, D is its discriminant, and Γ the gamma function.

We return now, as another example, to the Gauss sums

$$\sum_{v=0}^{n-1} e^{\frac{2\pi i}{n} v^2} \tag{$*$}$$

which Gauss first defined in connection with cyclotomic equations (see p. 30). After many unsuccessful attempts to determine the sign of $(*)$, Gauss succeeded by introducing and using the two series

$$f(x, m) = \sum_{v=0}^{\infty} (-1)^v (m, v) \quad \text{and} \quad F(x, m) = \sum_{v=0}^{\infty} x^{v/2}(m, v)$$

with

$$(m, v) = \frac{(1 - x^m)(1 - x^{m-1}) \cdots (1 - x^{m-v+1})}{(1 - x)(1 - x^2)(1 - x^v)}.$$

Though Gauss does not indicate how he found f and F, the definition of the two functions makes it obvious that there is some connection with the theory of elliptic functions. Gauss himself did not pursue the question, but Jacobi showed in 1817 that $(*)$ can be calculated within the theory of the linear transformations of the ϑ-function. Later, Kronecker proved the equivalence between the simplest case of linear transformations of the ϑ-functions with a particular formula for Gauss sums that goes back to Dirichlet. Kronecker's methods are analytic; in a very simple proof, he calculates $(*)$ with the help of Cauchy's integral theorem.

Already from Gauss's work it becomes clear that the sums play a central role in the various proofs of the law of quadratic reciprocity; later important applications are the calculation of the general class number formula and the summation of (Dirichlet's) L-series. One can generalize $(*)$ by considering characters of higher powers. This leads to even more difficult, still open questions.

As a last example, we mention a fragment in which Gauss proves the irreducibility of

$$x^3 + y^3 + z^3 = 0$$

over the field of third roots of unit. Gauss shows this—if one uses modern language—by first proving that the ring of integers in $Q(\omega)$, $\omega^3 = 1$, is Euclidean and so has unique factorization. Gauss's result includes Fermat's last theorem for the case $n = 3$. In another context, Gauss proved the theorem for $n = 5$, but he never occupied himself with it in a systematic way because he did not consider the question particularly fruitful for the further development of the theory of numbers.

The sketchy remarks of this interchapter must suffice as hints at the connection between Gauss's work and later, even present-day research. There is no doubt that many of Gauss's fragments have not yet been fully understood though, of course, it is arguable whether or not a modern understanding of what they mean is at all possible. There is little doubt that they contain much that can serve as an inspiration if one wants to find out how much one can learn in number theory by straightforward calculations and skillful manipulations; Gauss's work makes clear the concrete roots to much of the abstract apparatus that was developed by his successors.

Much of the contents of this interchapter is included in Rieger's essay in [Reichardt] on Gauss's arithmetical work, which will provide the interested reader with substantial additional material.

CHAPTER 4

The Return to Brunswick. Dissertation.
The Ceres Orbit

In the fall of 1798, Gauss returned to Brunswick, where he lived until 1807. It is obvious to us, as it must have been equally obvious to Gauss himself, that the coming years would be of critical importance to his career. It had to be seen whether Gauss was able to satisfy, by successful application of his talents, his own expectations and the expectations of his teachers and friends, particularly the Duke's. Up to now, Gauss's only publication was the announcement of the constructibility of the 17-gon; otherwise, his "works" consisted of a plethora of drafts and fragments, of unfinished essays and unpolished ideas of whose significance and meaning the author himself could not have been too sure. But it was not only professionally that Gauss had to prove himself—he was now 21 years old, returning to his native town where his parents still lived, and trying to secure an independent existence for himself, possibly even to establish a family and assume the responsibilities of an adult (see also p. 60). An important first step was to choose not to live in the house of his parents; soon after his arrival in Brunswick, Gauss gave instructions for his new address in one of his letters to Bolyai.[1]

It is not possible to claim that Gauss worked more, or harder, in these years than in the preceding or subsequent years at Göttingen. We know only that this second Brunswick period was uniquely fruitful and productive. It is also during these years that we can observe an enormous expansion of Gauss's scientific interests—for the first time he devotes himself systematically to questions of applied mathematics, specifically to theoretical and experimental astronomy.* An analogous development took place in the private sphere. Never in his life was Gauss more mobile and open than during this period, when he met many new people, started friendships, and made several important trips. Not without justification is it claimed that Gauss was never happier in his life than during the seven years of his second stay

---

* Gauss appears already to have shown some interest in astronomy when he was in Göttingen, but these first efforts cannot be compared with his at times total involvement starting from 1801.

in Brunswick.[2] At the close of this period came another incisive step: he met, courted and married Johanna Osthoff, the first of his two wives.

Even though these years were productive, full of exciting scientific discoveries and private satisfaction, life was strangely insecure, without definite prospects for the future. We notice, especially during the later years, manifestations of irritability and a certain general weariness, often triggered by the unavoidable commonplace chores. There was also a tedious conflict with the Collegium Carolinum, his former school, about the use of a certain valuable astronomical instrument—an unimportant, but quite revealing controversy to which we will return below.

At the time of his return to Brunswick, Gauss had no secure income and no concrete expectation of a teaching job (or inclination for one) at any of the academies in Brunswick or vicinity. Gauss's subsistence and immediate future depended on the benevolence and munificence of the Duke; on September 30, 1798, Gauss wrote the following to Bolyai:

I have reason to hope that the Duke will continue his assistance until I attain a secure position. I missed a certain lucrative one. There is a Russian emissary here whose two young and intellectually gifted daughters I was supposed to instruct in mathematics and astronomy. I was, however, too late, and a French emigré obtained the position.[3]

Early in January 1799, Gauss was in a position to inform Bolyai, then still in Göttingen, that the Duke had agreed to continue his stipend of 158 talers per annum. Until then, Gauss had lived on borrowed money, and as late as November he advised Bolyai to answer any curious questions by saying: "... that I have good though not quite certain prospects, which is in principle true" ("... dass ich gute wenn gleich noch nicht ganz bestimmte Aussichten habe, was ja auch im Grunde wahr ist"[4]).

Shortly before the Duke requested the submission of a doctoral dissertation, Gauss had been in touch with J. F. Pfaff, Professor of Mathematics at the University of Helmstedt.* Gauss's initial concern was use of the library. In the course of his studies he came to know Pfaff well, and stayed as his guest at his house for several weeks. Pfaff's contributions to differential geometry and the theory of partial differential equations are still remembered. He was a competent mathematician and a kind man.[5] Pfaff was Gauss's thesis advisor but we do not know whether he was at all scientifically involved. On June 16, 1799, even before the thesis was published, Gauss was awarded the title *Doctor Philosophiae* after the usual requirement of an oral examination ("defence"), certainly particularly tedious to Gauss, was dropped. The dissertation, the publication of which was financed by the Duke, appeared in August 1799 under the title *Demonstratio nova theorematis omnem functionem algebraicam rationalem integram unius variabilis in*

---

* Helmstedt lies approximately 25 miles east of Brunswick, very close to the Prussian border.

factores reales primi vel secundi gradus resolve posse; in a letter to Bolyai, Gauss described its contents in the following way.

The title describes the main objective of the paper quite well though I devote to it only about a third of the space.* The rest mainly contains history and criticisms of the works of other mathematicians (namely d' Alembert, Bougain-ville, Euler, de Foncenex, Lagrange, and the authors of compendia—the latter will presumably not be too happy) about the subject, together with diverse remarks about the shallowness of contemporary mathematics.[6]

Pfaff shared with Gauss an interest in the foundations of geometry, but it is mere speculation that the two discussed this topic.

Gauss's dissertation is about the fundamental theorem of algebra. The proof and discussion (see below) avoid the use of imaginary quantities though the work is analytic and geometric in nature; its underlying ideas are most suitably expressed in the complex domain. Like the law of quadratic reciprocity, the fundamental theorem of algebra was a recurring topic in Gauss's mathematical work—in fact, his last mathematical paper returned to it, this time explicitly using complex numbers.

We confine ourselves to a brief sketch of the contents of the dissertation. Volume X,2 of *G.W.* contains a very instructive essay by Ostrowski which explains the details of Gauss's line of thought and fills certain gaps in his argument. Let the polynomial $x^m + Ax^{m-1} + Bx^{m-2} + \ldots = 0$, $A, B, \ldots$ real, be given. The first step consists of the decomposition of the expression

$$X = x^m + Ax^{m-1} + Bx^{m-2} + \cdots$$

into its real and its imaginary parts; $X = T + iU$. Both curves are represented with the help of polar coordinates. A root exists if the two curves $U = 0$ and $T = 0$ intersect. The fundamental theorem follows from the separate investigation of the two curves. The idea of the proof is intuitively accessible; the proof itself is not rigorous according to our present standards, but easily superior to any earlier effort, specifically d'Alembert's, to which it is related.

Of the three other proofs which Gauss gave, the last is very similar to the first and will briefly be discussed in the context of its publication in 1849. The third proof, published in 1816 in the paper "Theorematis de resolubilitate functionum algebraicarum integrarum in factores reales demonstratio tertia" is genuinely analytic. Gauss uses Cauchy's integral theorem in an implicit way by considering, for the polynomial f, the expression

$$\int_{|X|=r} \frac{df}{f} \qquad (*)$$

$(*)$ vanishes if X does not have a root. This leads to a contradiction. Our

---

* In the original edition, the dissertation is 80 pages long, but it takes up only 30 pages in the format of *G.W.*

summary is more direct than Gauss's proof because Gauss avoids complex numbers and explicit geometric constructions. Instead, a real double integral is used.

The second proof, "Demonstratio nova altera theorematis omnem functionem algebraicam rationalem integram unius variabilis in factores reales primi vel secundi gradus resolvi posse", from 1815, uses algebraic properties (of the symmetric functions) and a differential equation between the initial polynomial and its discriminant. Gauss's idea is related to the modern, abstract approach used to prove the existence of a splitting field and to show that it is contained in the field of complex numbers.

In his booklet *Gauss zum Gedächtnis*, Sartorius, Gauss's first biographer, claims that Gauss's interest in geometry dates only from comparatively late in his life. If this is based on any authentic statement it could be the result of a misunderstanding and perhaps refer to geometry as a subject of mathematics in its own right rather than to it as a mathematical tool and "way of thinking". We have certainly no reason to trust Sartorius' judgment too much. He was a geologist without deeper understanding of Gauss's mathematical work.

In the same way as he used numerical examples in his heuristic considerations, Gauss used geometric visualizations quite frequently. An example is the illustration below which is similar to several others in Gauss's fragments dealing with the theory of biquadratic and cubic residues. The different compartments are discontinuity areas of a certain group of linear transformations. (For more information, see pp. 18–20 in Vol. 8 of *G.W.*)

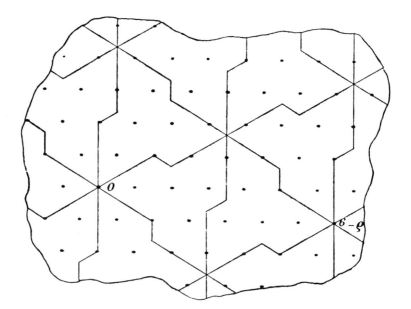

Gauss was among the first mathematicians to use d'Argand's visualization of the complex numbers in the 2-dimensional plane. He probably preceded d'Argand but there were others who independently came up with the same idea.[7]

From the diary we know that Gauss's interest in mathematical and observational astronomy goes back to the time when he was a student in Göttingen, if not to his years at the Carolineum. In Göttingen, Gauss was, as mentioned before, in contact with the astronomy professor Seyffer, but we do not know whether he had occasion to make any observations. By studying the classical literature he learnt about the main problem of mathematical astronomy, the computation of the orbits of celestial bodies from inaccurate and scarce observations. Gauss also studied the theory of the Moon. The years 1798–1800 were largely occupied by the completion of *Disqu. Arithm.* and the dissertation, but as early as April 1799 Gauss expressed in a letter to Bolyai his intention to visit the astronomer Zach in his observatory Seeberg near Gotha. The visit did not in fact take place at that time though permission to go abroad had already been granted by the Duke; instead, Gauss went several years later. The Baron (*Freiherr*) Zach was one of the best-known German astronomers, in charge of a very modern observatory, fitted out by the liberality of his prince. Seeberg was the center of astronomical research in Germany.

The years around 1800 are important in the history of astronomy: for observational astronomy because the technological progress and the refinement of the optical instruments, together with the systematic accumulation of observations, led to the compilation of the first reliable sky maps; and for theoretical astronomy because the discovery of the outer planets (Uranus 1781, Neptune 1846; Pluto, however, not until 1930) supplied the data needed for precise calculation of the perturbations of the planets. The mathematical tools for this difficult and complicated task had been available earlier, but only now did one have the necessary data and facility with analytic techniques to deal with them. Without doubt, Gauss felt attracted by the practical and technical aspects of observational astronomy. Zach discussed this in his letter of February 21, 1802, when he tried to persuade Gauss to devote himself to computational and theoretical questions, where his genius was unique, rather than lose himself in the tiresome efforts of observation.* In addition to being director of the Seeberg observatory, Zach was the editor of *Monatliche Correspondenz*, at that time the main German astronomical periodical. In June 1801, Zach published the orbital positions of a new planet, Ceres (Ferdinandea) which had been discovered by the Italian astronomer G. Piazzi

---

* In all fairness, it should be mentioned that it was also Zach who reassured Gauss about his shortsightedness. The question (i.e., the discussion of the optical system consisting of myopic eye and telescope) recurs in Gauss's later correspondence, and Gauss does not seem to have been able to find a conclusive answer to what the concrete effects of the observer's shortsightedness are.

on January 1, 1801. Only 9° of its orbit had been observed before Ceres vanished on February 11 of the same year in "the shadow of the Sun". Its existence had been suggested by Bode's law, and astronomers all over Europe prepared themselves for the rediscovery of Ceres when it was expected to reemerge at the end of 1801 or early in 1802. Zach published several forecasts of the prospective orbit, among them one of his own and one by Gauss, the latter in the September issue of *Monatliche Correspondenz*. Though *Disqu. Arithm.* had only been published on the 29th of the same month, Gauss already enjoyed a certain fame as an exceptional mathematician, not, however, as an astronomer. Still, his orbit was taken seriously, though it was quite different from the others and substantially expanded the area of the sky which had to be searched. First Zach, in the night of December 7, 1801, and then Gauss's later friend W. Olbers on New Year's Eve managed to locate the new planet at positions very close to the ones predicted by Gauss. The result, published in the February issue of *Monatliche Correspondenz*, made Gauss a European celebrity—this a consequence of the popular appeal

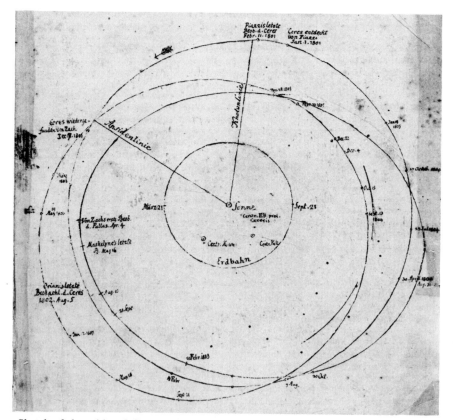

Sketch of the orbits of Ceres and Pallas (nachlaß Gauß, Handb. 4). Courtesy of Universitätsbibliothek Göttingen.

which astronomy has always enjoyed and of the international cooperation which existed among astronomers because of their need for reliable data and the paucity of competent observers and modern observatories.*

This success brought Gauss many honors, among them an invitation to St. Petersburg to become director of the local observatory.[8] Russia had a tradition of inviting eminent foreigners to positions at the local scientific institutions. Euler is the best-known example, and Gauss himself had just been made Corresponding Member of the Academy on account of *Disqu. Arithm.*—with a subsequent confirmation and improvement of his position in Brunswick. It also made Gauss a member of the small group of serious German astronomers with whom he henceforth communicated on an equal and personal footing. Here we see for the first time how Gauss, although primarily a mathematician, finds respect, a circle of friends, and a sphere of scientific cooperation, exchange, and influence among nonmathematicians.

In June 1802, Gauss was for three weeks the guest in Bremen of Dr. Olbers, a physician with a good reputation as a theoretical and observing astronomer. Olbers had just discovered Pallas, the second of the small planets. In 1803, Gauss finally met Zach and helped him in some geodetic work. In the same year, Gauss again went to Bremen for another visit with Olbers.

When he was a student Gauss was scientifically very isolated. Astronomy, as we saw, changed all this, not only on the scientific but also on the personal level. Why this was so is not difficult to see. Cooperation and scientific exchange are much more important in astronomy than in mathematics, and Gauss had to participate in these discussions for his own benefit. His theoretical efforts in astronomy naturally led him to the empirical, observational side of the subject, an occupation which clearly had much appeal for him. As in mathematics, he could work by himself, and he may have found the relaxing concreteness of the observations a welcome counterweight to his abstract theoretical work. Gauss became a regular observer in Brunswick despite the limited possibilities there; among astronomers, he quickly gained in authority, not only because of his modesty and his theoretical successes, but also because of his efficiency and preciseness as an observer. One of his specialties, to be discussed later in more detail, was the design and improvement of astronomical instruments.

Brunswick did not offer Gauss much in this regard, but in 1802–1803 serious discussions were started on building a small observatory for him.[9] This was in line with other efforts of the Brunswick government: everything possible was done to create favorable conditions for the famous young scientist and to keep him in his native town.

In 1802, when Gauss had not yet decided about the invitation from St. Petersburg, Olbers alerted his friend von Heeren, a professor in Göttingen

---

* As early as 1796, an "International Meeting of Astronomers" took place at the Seeberg observatory.

and advisor to the Hanoverian government. Olbers wanted to make sure that Gauss would not leave Germany and asked whether Gauss could not be made director of the newly projected observatory in Göttingen. He wrote:

> ... although mathematics and astronomy are not your fields, you know, dear friend, the fame that was gained by Dr. Gauss in Brunswick. It is fully deserved; this young man of 25 years is the first of his mathematical contemporaries. I believe I am quite a competent judge because not only have I read his works, but I have also been his trusted correspondent since the beginning of the year. His knowledge, his extraordinary dexterity in the analytical and astronomical calculus, his indefatigable activity and industriousness, his uncomparable genius aroused my highest admiration which has even been increasing, the more of his ideas he has been communicating to me in our correspondence. He loves astronomy enthusiastically, principally practical astronomy, little opportunity—for lack of instruments—as he has to exercise it. He would definitely not like to be a teacher of mathematics: he desires to be an astronomer at some observatory, so that he could divide his time between observations and his deep investigations for the sake of the expansion of knowledge ...[10]

We do not know of any official reaction to this letter, either from Göttingen or from Hanover; the matter was allowed to rest at the time. Because of the improvement of the conditions in Brunswick, Gauss decided to stay. Olbers, still concerned, took up the matter again a few years later, with a better response and more success.

It is difficult to predict planetary or similar motions from a small number of observations (at least three), because six equations in six unknowns have to be solved. These equations are so complicated that one has to approximate the solutions; it is not possible to calculate them accurately and to present them in closed form. The first step in such an approximation is the statement of a probable or possible orbit, the second, much more difficult one, the stepwise correction of the orbit. Basically, there are three types of orbital curves possible, elliptic, parabolic and hyperbolic.

Numerous techniques existed before Gauss—they had been used and refined in the evaluation of the motions of many comets—also one planet, Uranus (discovered in 1781 by Herschel), had been treated thus. Uranus was particularly simple because the initial assumption of a circular orbit around the sun was quite good and did not lead to any gross distorting errors. In addition, there existed in this case a large body of observations, including older ones by Flamsteed and Th. Mayer sr., which made it easy to compute the precise orbit. For Ceres, there were only Piazzi's 41 days of observation; moreover, its orbit turned out to have a considerable eccentricity, which made the circular hypothesis, used by Olbers for his prediction, quite inefficient. Gauss differed from his contemporaries by avoiding any arbitrary asumption for the initial orbit; his ellipse was based on the available observations only, without any additional, hidden or open, hypotheses. Much later, in "Theoria Motus . . ." (1809) Gauss explained a refined, quite different approach; his original computation of the Ceres orbit was largely based on heuristic considerations.

It is interesting to study in this context the paper "Summarische Übersicht der zur Bestimmung der Bahnen der beiden neuen Hauptplaneten ange-wandten Methoden" ("Summary survey of the methods used for the deter-mination of the orbits of the two new main planets"; *G.W.* Vol. VI). It was published in 1809, but written considerably earlier. Characteristically, even here Gauss could not refrain from refining and "mathematizing" his original method, which appears to have consisted largely of interpolations and step-wise improvements. This is surprising, because "Summarische Übersicht. . ." was the first explanation of his methods which Gauss published; it came eight years after he had first applied them and in response to numerous requests to explain his successful new technique.

Naturally, Gauss's initial orbit had to be further improved after Ceres had been rediscovered, and we see Gauss in lively correspondence with leading observers, among them Olbers, with whom a lifelong correspondence now started. It is an exchange in which Gauss incessantly computes new orbits which in turn lead to new observations. Gauss's skill for calculations was an essential factor in his success, combined with his strict application of the method of least squares (which will be discussed later in a different context).*

---

* To some degree, Gauss's success can be explained by the fact that he had previously studied the theory of the Moon, in which similar methods are used, especially for the approximation of the elements of the orbit by (finite parts of) Taylor and trigonometric series.

CHAPTER 5

Marriage. Later Brunswick Years

On October 9, 1805, Gauss married Johanna Osthoff, a tanner's daughter to whom he had become engaged the previous year. Johanna was three years younger than her husband; her family had been friendly with Gauss's mother.[1] The groom seems to have known his future wife as a child but we know nothing of the specific circumstances which led to the marriage. Neither do we know much about the young couple's private life, but one can get some insight into the marriage from the correspondence. Several letters have been preserved; most were written during Gauss's trip to Bremen in the summer of 1807, shortly before the move to Göttingen. Like most of Gauss's family correspondence, the letters are reprinted in *C. F. Gauss und die Seinen* (see Bibliography). We quote from Gauss's letter of June 27:

Yesterday noon, Friday, I arrived here after a very inconvenient trip: my first letter belongs to you, dear Hannchen. The terrible weather to which I was exposed without interruption, the rain that went through my chenille coat, bath robe, two frocks, and shirt made me thoroughly dislike this type of travelling: fortunately, it did not do me any harm and I came away with some temporary discomfort. I did not encounter any other difficulties: all the roads here are perfectly safe, and nobody asked for my passport.

Olbers is not quite well, he has shingles on one cheek and is not allowed to expose himself to the fresh air; apart from this, he and his family are fine, all send you their cordial greetings. Up to now (9 o'clock in the morning) I have not met any new people besides Dr Focke and his little Wilhelm, a lovely and very healthy child of two years, at 10 months he could already walk quite safely. I hope to see Bessel today. Olbers is very encouraging with regard to the Göttingen matter; he thinks I can totally rely on the certainty of the engagement. If it materializes I will probably try to start there not at Michaelmas, but rather in the course of the fall or around New Year, because I will not be able to lecture next winter. In any event, there will be no changes with our apartment.

I waited in vain for a letter from you today: I hope everything is fine and you are all well. Please write soon how you all are, whether your good mother recovered from her shingles (Olbers is better, without any treatment, except that he takes good care of himself), how our dear Joseph is, how satisfied you are with his new nurse. Did you pay for Mr. Mengen's wine—perhaps you do not know any more how much it is, I think it is 1 taler, 14 groschen, and 3 pfennigs.[2]

Gauss's two letters to Bolyai, of June 28 and November 25, 1806, are among the most open and emotional letters he ever wrote. Gauss was concerned about his future, but also about life with his young wife who was such a different person—intelligent and sweet but also inexperienced and not well educated. Two quotes from his letter of November 25 illustrate the situation: "Life appears to me like spring with its shining colors," but also, "A one-sided happiness is no happiness at all." No portrait of Johanna Gauss has been preserved; we only know that her daughter Minna, later the wife of the orientalist Ewald, looked very much like her. Gauss described his bride in his letter to Bolyai of June 28, 1804:

The beautiful face of a madonna, a mirror of peace of mind and health, tender, somewhat fanciful eyes, a blameless figure—this is one thing; a bright mind and an educated language—this is another; but the quiet, serene, modest and chaste soul of an angel who can do no harm to any creature—that is the best.[3]

In addition to the one quoted above, a few more letters between Gauss and his first wife are known; it appears that their love was direct and open, expressing a deep but conventional affection. Though we do not know much of Johanna, she strikes the reader as quietly assertive, endowed with an inner security which allowed her to retain her independence and personality. She was not totally overshadowed and overwhelmed by the genius of her husband, despite her inferior education and comparative inexperience.

There are a couple of portraits of Gauss from this period; his appearance seems already to be quite similar to the familiar later portraits. Gauss was of average height, about 5ft 2 in., stocky and "of characteristic low-German features". The clear and penetrating blue eyes are mentioned often; in the later portraits the jutting forehead is quite prominent, but this may not have been quite so conspicuous in his youth. Gauss was muscular and strong, certainly not the proverbial debilitated scholar.

December 1804 marks the beginning of the correspondence with Friedrich Wilhelm Bessel, then 20 years old and working as an apprentice for one of the great merchants in Bremen. His interest in astronomy brought him in touch with Olbers, who in turn recommended the young, self-taught man to Gauss, for whom he analyzed observational data. This was an important and responsible task in the pre-computer age; the analysis of observations, their *reduction*, was an essential and time-consuming part of astronomy. Even Gauss, with all his computational skill, was thankful for the help.

Bessel developed after a short time into an equal partner in their astronomical discussions; we shall see that this change was not without consequences in their personal relations. But during Gauss's years in Brunswick,

Bessel was still the docile and obedient pupil; Gauss gave him much encouragement and discussed with him interesting questions such as the determination of the orbits of certain historical comets. Gauss and Bessel had a very extensive and substantial correspondence, but for nearly 20 years, between 1807 and 1825, they did not meet in person though there had been several opportunities. The letters are not confined to astronomical matters, where they are of the nature of a lively and informal discussion. They also cover mathematical topics—where Gauss is more magisterial—and some private and personal affairs. Examples are Gauss's gladly undertaken efforts to have Bessel and his younger brother exempted from French military service, or to take care that Bessel received a doctoral degree from Göttingen without thesis or examination after he had been made director of the observatory at Königsberg in Prussia.[4] The correspondence with Bessel is the scientifically most interesting and informative part of Gauss's correspondence, but he exchanged far more letters with Olbers, Schumacher, and Gerling. Gauss appears to have been an indefatigable correspondent, but this was not unusual for the time. Because there were so few active and sympathetic scientists and so many obstacles to direct personal exchange, letters were a most important and indispensable instrument of scientific progress. The modern reader may be surprised to hear that, at least within Germany, the regular mail was both reliable and quick; even from places as remote as Königsberg (in Eastern Prussia) or Munich (in Bavaria), a letter to Göttingen took no longer than a week during times of peace. The system was even more efficient in the 1830s and 1840s, when Gerling's letters from Marburg used to arrive in Göttingen overnight.

The dominant theoretical theme of the later Brunswick years was the calculation of the perturbations of the orbits of Pallas and Ceres which were induced by the mass of Jupiter. Many scrapbooks with computations are preserved; implicitly, they contain the theory which Gauss intended to work out later. For reasons which we shall see presently, Pallas was particularly obstinate. An interesting feature of these problems is that they were quite closely connected with Gauss's work in pure mathematics, most conspicuously with his research on elliptic integrals, and the hypergeometric function. For more details, one should consult Vols. III, VI, and VII of Gauss's collected works.[5]

Gauss essentially used two different methods for his computations of the irregularities of planetary motions. One consists of the analytic expansion of the elements of a perturbation. Only the first elements of the resulting infinite series are taken into account and used. This, of course, was not an original idea with Gauss. Laplace, among others, proceeded similarly. But Gauss was more efficient than his predecessors and competitors, mainly because of his familiarity with a great number of infinite series and his skills in manipulating them. This analytic method did not work for complicated or very slowly converging series; instead, Gauss invented and used numerical

integration. Though always feasible, the method was very time consuming—
every observation was a separate problem. Gauss was not one-sided or
categorical in his preferences and chose his methods according to the situation. The analytic methods were satisfactory for Ceres, but even then Gauss
expanded the perturbations into trigonometric series which he integrated
numerically with the help of tables. Gauss's techniques are very similar to
the work of Fourier and anticipate it to a considerable degree.

For the mathematical astronomer, the determination of the perturbations
of the small planets was the most important task of the time. It obviously
was a congenial task for Gauss, though the Pallas perturbations were too
much even for him. One of the most interesting results which one could
expect from these calculations was a precise determination of the mass of
Jupiter; in a letter of June 25, 1802, Gauss wrote to Olbers: "By the way, I
believe that, after a few revolutions, Pallas will be most suitable for determining the mass of Jupiter" ("Übrigens glaube ich, dass die Pallas nach
einigen Umläufen das beste Mittel sein wird, die Masse des Jupiter zu bestim-
men")—a prediction which turned out to be overly optimistic.

Despite his preoccupation with astronomy, Gauss was actively interested
in several other areas of natural science: gravity experiments, in order to
investigate the rotation of the earth; the determination of geographic longi-
tudes, under the direction of Zach (by acoustic and optical signals);[6] and
astronomical observations. We have already mentioned that there was no
proper observatory in Brunswick, but on good nights Gauss was able to
locate Pallas and Ceres. Of considerable interest were some of the more
visible comets, particularly one which appeared in 1805; it was this comet
which he discussed at length with the young Bessel, who investigated whether
it could be identical with the well-documented comet of 1772. (In fact it was
not.)

The showpiece, unfortunately not much more, of the instruments which
were at Gauss's disposal in Brunswick, was a modern reflecting telescope.
Gauss had obtained the necessary funds in 1804 and subsequently purchased
it with the help of his future colleague, the astronomer Harding, a skilled
and knowledgeable observer. The reflector turned out to be defective and
had to be adjusted several times. Gauss consequently was never able to use
the instrument while he worked in Brunswick and when it became known
that Gauss would accept a position in Göttingen, the professor of astronomy
at the Carolineum officially requested the instrument. Gauss wrote a long
opinion for the minister, explaining the history of the purchase and the
defects and virtues of the instrument. It concluded with the recommendation
that the telescope be handed over to Pfaff as the best-qualified potential
user in the duchy. It appears that the instrument nevertheless stayed in
Brunswick and was incorporated in the collection of the Collegium Carolinum. Gauss, very disappointed about this turn of events, did not hide his
disapproval. We quote from a letter to Olbers which was written immediately

before his departure to Göttingen, on October 29, 1807:

The reflector of the 10-ft telescope which was returned to me appears to be quite good now; lack of space does not allow me to use it much; also, I have neither the inclination nor the time to center it precisely. It is easily possible that the instrument will fall into very bad hands after I leave.[7]

The imperfections of the reflector prompted Gauss to concern himself with what he called his "dioptrical investigations", that is, the investigation of systems of optical lenses and their theoretical and real limitations. This area remained of continuing interest to him; his contributions are considerable and will be summarized later.

His discussions with Olbers and, slightly later, with Bessel, convinced Gauss quickly of the need for more accuracy, strictness, and fidelity in astronomical observations. This applied to the observations themselves as well as to their evaluation in a scientifically (i.e., mathematically) sound way. Gauss was a relative newcomer to the field; still, this was an area in which he could excel and set standards, despite the constraints of the Brunswick and early Göttingen years when he had no modern instruments. Later, Gauss was to apply the same strict standards and the same approach to the other areas of the experimental sciences in which he worked. He felt very strongly about this and openly criticized the sloppy work of earlier observers, particularly geodesists. Throughout these reductions, his most useful theoretical tool was the method of least squares, the same technique which had proved its value in the determination of the Ceres orbit. Today, we know that Gauss and Legendre can independently claim credit for the method of least squares (see pp. 138ff.), but even Gauss did not at the time view it as a theoretically very interesting or important method. Initially, he was convinced that Tobias Mayer sr., his predecessor at the Göttingen observatory, had known and used it.[8] It was only later, after Gauss had discovered its probabilistic motivation, that the method became of intrinsic interest to him and developed into an important element of what may be called Gauss's natural philosophy.

In October 1805, in his correspondence with Olbers, Gauss first mentioned his intention of explaining his methods in theoretical astronomy in a comprehensive treatise. This book was finally published in 1809; it is the famous *Theoria Motus Corporum Coelestium in Sectionibus Conicis Solem Ambientium*. The completion of this manuscript was the biggest project of the later Brunswick years; its publication was an event that all astronomers looked forward to with considerable expectation. Its contents will be summarized below.

There is a last, though minor, point which should be mentioned in a summary of Gauss's scientific work during the second Brunswick period. In the years 1804 and 1805, he had some correspondence about number-theoretic problems with a then unknown French mathematician. We quote twice from the correspondence with Olbers, first from a letter of December 7, 1804:

Recently I had the pleasure to receive a letter from LeBlanc, a young geometer in Paris, who made himself enthusiastically familiar with higher mathematics and showed how deeply he penetrated into my Disqu. Arithm. . . .[9]

The second quotation is more than two years later (March 24, 1807):

Recently, I was greatly surprised on account of my Disqu. Arithm. Did I not repeatedly write you of a correspondent in Paris, one M. LeBlanc, who had perfectly understood all my investigations? This LeBlanc recently explained himself to me. You will certainly be as surprised as I was when you hear that LeBlanc is the assumed name of a young woman, Sophie Germain.[10]

One of the later proofs of the law of quadratic reciprocity is connected to an idea of Sophie Germain;[11] her name is remembered in number theory because she was the first to find the solution for certain special cases of Fermat's last theorem.[12] This last remark again illustrates the interest which Gauss retained in pure mathematics during these years which are otherwise over-shadowed by his astronomical work. In our earlier review of the number-theoretic work we saw that he not only published several papers that continued the ideas of *Disqu. Arithm.*, but he also did new research and discovered some new results, without, however, attaining his original, much more far-reaching objectives. All his systematic efforts were directed towards astronomy, for the reasons which were pointed out above. They essentially are, in arbitrary sequence, "scientific" astronomical interest, the superior possibilities for scientific and personal exchange which astronomy offered, and the expectation it would be easier to find a suitable and secure position as a astronomer. We can assume that this last motive was reinforced by his marriage and the birth of his first son in 1806.

When accepting the appointment in Göttingen Gauss wrote that he never considered the position in Brunswick as more than temporary,[13] but he preferred it, as late as 1804, to offers from the academy in St. Petersburg and the University of Landshut (Bavaria). The negotiations with Göttingen entered a serious phase in 1804, two years after Olbers' first letter, and were successfully concluded in 1805, i.e., before the political revolutions of 1806. Olbers, the indefatigable advocate of Göttingen (and of Germany, for that matter), conducted the negotiations on Gauss's behalf. The decisive reasons in Göttingen's favor were the firm commitment of the administration to erect a new observatory, the presence of the experienced and skillful observer C. L. Harding (the discoverer of the small planet Juno) as Gauss's assistant,[14] and the fact that Gauss would only loosely be connected with the university. This gave him relative freedom from lecturing and from participation in the administrative affairs of the university. An additional—negative—reason was the apparent vagueness of the plans for the projected observatory in Brunswick.

The Political Scene in Germany, 1789–1848

Even before Gauss arrived in Göttingen, his decision to go there was vindicated by political developments. Soon after 1789, a series of short wars erupted between the "Roman Empire" and France, led since 1799 by the seemingly invincible Napoleon. These wars were concluded by the defeat of the German states between 1805 and 1807. In a last effort, the Prussian government sent Duke Ferdinand of Brunswick to St. Petersburg in 1806; his embassy was to sound out the possibilities of a coalition against France. The desired alliance with Russia did not materialize, and Prussia decided to proceed on its own to stop the French advance in central Europe.

Ferdinand of Brunswick, Gauss's Duke, was one of the most famous soldiers of his age. He could look back on a military career of nearly 50 years, originally as a Prussian general in the glorious service of King Frederick the Great.[1] Though he was now over 70 years old, the timid and backward-looking Prussian cabinet made him generalissimo of the Prussian army. The first major battle decided the war; Prussia was beaten at Jena and Auerstaedt,[2] the Duke was mortally wounded. He died a few days later in Altona near Hamburg. His party, taking refuge from Napoleon, went through Brunswick, and we know that Gauss, early one morning, was stirred from his sleep by the melancholic rattle of the coaches which left the city through the Hamburg gate.[3]

The political situation in Göttingen was quite complicated. The Kingdom of Hanover, under the English crown since the ascension of the House of Hanover to the British throne nearly a hundred years earlier, had been occupied by Prussia immediately before the Franco-Prussian war; after Prussia's fall, England could not protect it, and the country came under direct French dominion. The short trip from Brunswick to Göttingen, taken by Gauss 13 months after the fateful battle, amounted to a journey from the age of feudalism into the bourgeois 19th century. Göttingen was incorporated in the new, French-dominated Kingdom of Westphalia, and the ease with which Gauss was allowed to make this transition, both concretely and symbolically, was quite exceptional, just as his easy relations with the Duke had been untypical. The key historical events through which Gauss lived passed him by without deeply disturbing or upsetting him: *Sturm und Drang*, the French Revolution, the rebellion against Napoleon, and the revolutions of 1830 and 1848 did not really affect Gauss: they caused him no major inconvenience and he was not a victim of the oppression which helped to bring them about—Gauss never had reason to salute the dawn of a new morning and hope for momentous changes in a new age.

We saw how kind absolutism had been to Gauss, but the much-hated French-controlled government of Westphalia was no less benevolent towards the sciences and towards him. More unpleasant was that the government was quite inefficient and incompetent; its progressive deterioration[4] showed itself in the tardiness with which the construction of the new observatory proceeded and in occasional delays in the payment of Gauss's salary. There is an ironic marginal note which should be mentioned: one of Gauss's articles had to be delayed because all the disposable printing capacity and paper had been used up by the publication of yet another volume of the collected works of the "polyhistor" Joh. von Müller, then Secretary of Education and Culture in the Westphalian government and proud author of more than 200 learned tomes. Because the political situation was so unsettled when he took office as a civil servant, Gauss was never asked to take the customary oath—an omission which seems to have filled him with a measure of gleeful satisfaction, more because it was so curious than because of any legal consequences.[5]

When the old Kingdom of Hanover was restored in 1814, it was probably the most conservative state in Germany; still, Gauss could feel comfortable because of the official appreciation astronomy enjoyed—again, the English influence is obvious. The funding for the observatory was liberal, and Gauss obtained all the instruments and all the help he could reasonably have expected. His conflict with society took place in the private and personal sphere, and is expressed, for example, in his dissatisfaction at the slowness of the military career of his son Joseph.[6]

The world which Gauss had learnt to consider as normal collapsed in 1806; nevertheless, the impact which the political developments had on his own fate was not strong enough to force Gauss into revising his basic attitudes. Probably his most unpleasant experience was service for the unloved Kingdom of Westphalia, but even this was mitigated by the unquestioned good intentions of its inept government and by the personal honors which were bestowed on Gauss—in a curious anachronism, he was knighted and could call himself "Ritter von Gauss" for a few years.[7]

The first 30 years of Gauss's life have already supplied us with some of the information which will help us to understand his political attitudes. His origins and his career were determined by the social and political forces of the 18th, not of the 19th century; the prince to whom he owed much of his advancement was a benevolent despot in the style of Louis XIV. Duke Ferdinand of Brunswick was efficient and far-sighted by the standards of his time (though he did not disdain to improve his budget by the sale of

soldiers for service in North America). He was an educated man, philosophically a physiocrat and a correspondent of Diderot and the older Mirabeau.
One of his early advisors was the *Freiherr* Hardenberg, who was later instrumental in the introduction of radical reforms in Prussia during the years
1809–1813. One of the Duke's most memorable actions was the employment
of the writer G. E. Lessing as librarian in Wolfenbüttel.[8]

Nevertheless, the Duke's enlightened absolutism did not do much good
in a small country like Brunswick. It did nothing to reduce, it even widened
the gap between the ascending third estate—the future bourgeoisie—and the
deprived majority of apprentices, farm hands, etc.* As in most other German
states, productivity and welfare did not increase in Brunswick during the
18th century. The rise of the middle class took place at the expense of other
social groups.[9] The system itself had lost some of its old rigidity—Gauss's
own family provides a good example. Gauss himself accepted this situation
and never considered nonconformist or radical social ideas. He tried to
advance his own children according to the rules of the current order, as,
for instance, when he tried to force his son Eugen to become a lawyer, a
profession which socially would have been most acceptable.[10]

Our way of seeing the 18th century has been determined by the course
of the political and national history of Germany during the 19th century.
There are two radically different approaches to German history of the last
100 years, one progressive and liberal, the other conservative; the origins of
both can be traced back to the 18th century.

The liberal point of view sees the 19th century in the following way. The
old feudal order broke down, its weakness exposed in the conflict with the
progressive popular forces of the French Revolution. The consequence of
this development was a temporary coalition between the democratic national
movements in Prussia (and the other German states) and the conservative
feudal forces which were fearful of a revolution in Germany and a complete
loss of power. Between 1813 and 1815, this coalition liberated Germany from
the foreign yoke by defeating Napoleon. By then, Napoleon himself had
become a conservative, even a reactionary historical force. The time between
1815 and 1848 is characterized by a revival of the dynastic and conservative
"party" in Germany. The ensuing tensions led to the abortive revolution
of 1848/49. Because the intervening years had been so well used by the con-
servatives, they succeeded, against the "real" historical development, in the
defeat of the revolutionary liberal forces. The result was the eventual creation
of a backward-looking and aggressive German national state under the
leadership of Prussia.[11]

The conservative interpretation stresses the national component in the
uprising against Napoleon and sees the revolution of 1848/49 as the revolt
of a few malcontented intellectuals, without the support of the people and

---

* This development was typical for Germany at the time. For a detailed analysis, see
[Biedermann].

triggered not by the situation in Germany, but by the events in France. This is considered to be the reason for its quick suppression and the ineffectiveness of the liberal parliamentary assembly (*Paulskirchenparlament*). The real goal of the wars against Napoleon and France was only attained in 1871 under Prussian leadership, after national wars against Denmark, Austria, and France.

From a liberal viewpoint, the unfulfilled dream of a democratic and unified (and consequently strong and peaceful) Germany was the real aim of the Napoleonic wars, which are interpreted as a popular rather than a national uprising.

Neither of these approaches seems to do full justice to the historical development. The origins of the different interpretations (and, of course, of the events they describe) go back to the 18th century. At that time, even many of the central concepts of the discussion, like *Nationalgefühl und -bewusstsein* (national feeling and consciousness), had a different meaning. A concrete national sentiment oriented towards a unified German nation did not exist. The political organization was that of the "Holy Roman Empire of German Nation," an utterly lifeless fossil from the Middle Ages from which no change, no exciting initiatives could be expected. The old order was more a barrier on the road towards modernization and progress than the precursor of a new, unified Germany. The decentralization of Germany into several hundred semi-independent states had progressed so far that the mere idea of unification appeared to be an empty and obsolete phantasm.

Such a description of the political and national situation is correct, but it is incomplete in an essential way. Apart from any immediate political considerations, there were other statements of a totally different character. They assumed and even proclaimed the reality of an abstract notion of Germany and even of a German national character. Occasionally, such ideas were expressed with much self-assertion and a national pride which cannot easily be understood. The two main outlets for feelings of this kind were in contemporary literature and in historical study of certain political institutions. In the literature of the 18th century we see a strong trend towards cultural independence (from France), expressing itself in movements for keeping the language free from foreign (i.e., French) words and in the adoption of "German" meters in poetry.[12] Politically, it was the revival of the old institutions of popular representation, remnants from the late Middle Ages, that was emphasized; French absolutism, very popular with the petty rulers of the small German principalities, was rejected as "un-German" and incompatible with the old love of liberty of the Germans. This innate love of liberty was used as the reason why the Germans could not form or tolerate a strong national government.[13] In the works of contemporary journalists and historians* one finds many expressions of the conviction that all that was necessary

---

* For example, those of J. J. Moeser (1720–1794), a journalist, lawyer, and historian, with a background and convictions quite similar to Gauss's. Moeser came from Osnabrück, a Catholic enclave in an otherwise Protestant area of northern Germany.

for an alleviation of the present miserable state of German affairs was the restoration of the old order which had so well taken care both of the liberty of the Germans and of their temporal welfare.* Thus, it seemed quite natural to look backwards, and many of the progressive ideas imported from France by unruly intellectuals and adopted by sections of the slowly awakening middle class seemed strangely unnatural and so not applicable.

These two different strata of "liberal" thought can be discerned in the political thinking of Germany throughout the first half (if not longer) of the 19th century, both of them sharing the same slogans, like liberty, political freedom, and unity, but with different meanings and separated by completely different experiences and expectations. Many of the representatives of the liberal Germany personified this conflict, among them the poet Ludwig Uhland who was a radical deputy in the revolutionary parliament of 1848/49. In many of his poems, Uhland celebrates and praises *das alte gute Recht*, "the old good right" of the Germans.[14]

It is obvious that this emphasis on the past can be interpreted as conservative, but such an interpretation would be quite ambiguous and misleading. What "conservative" really meant as late as 1814 is ironically illustrated by the attitude of the Count Münster during the negotiations about the restoration of the office and the title of German Emperor. Count Münster was for many years the chief minister of the Kingdom of Hanover, for all intents and purposes its vice-regent. He argued it would be foolish to reestablish the title[†] because the old rights of the Germans, supposedly to be restored after the liberation from the French yoke, would leave the emperor without real power. This point of view prevailed, presumably for reasons different from Münster's original ones. Only gradually did German conservatism attain its national component in the course of the 19th century, undergoing a mutation and changing into the ideology which we today would call conservative; the right-wing politician, clamoring for military strength, extolling the virtues of the Germans and of Germany, and wearing an old-fashioned supposedly national costume is an invention of a later age and would have been inconceivable 150 years ago.

---

* As a foreign source, I quote Gibbon (Chap. XLIX of the *Decline and Fall*):

But the Italian cities and the French vassals were divided and destroyed, while the union of the Germans has produced, under the name of an empire, a great system of a federative republic. In the frequent and at last perpetual institution of diets, a national spirit was kept alive, and the powers of a common legislature are still exercised by the three branches or colleges of the electors, the princes, and the free and Imperial cities of Germany.

This description is certainly flattering and was anachronistic even then. It is an expression of the feelings of the time; Gibbon wrote this when Gauss was a child.

† Francis II had resigned the title in 1806, after Napoleon had "mediatized" and "secularized" the petty principalities and the independent possessions of the Church; Germany was reduced to around 35 medium-sized states. Francis, accustomed to the imperial dignity, subsequently assumed the new litle "Emperor of Austria".

We have gone into so much detail so as to avoid some possible (and common) misunderstandings of Gauss's political opinions and reactions. Classically, Gauss is seen from a liberal perspective, a view that has dominated much of the historical literature and appears to be the more enlightened approach. Gauss's origins were, as we saw, quite different from those of the classes which developed and identified with political liberalism. When Gauss entered the middle class he adopted only some of its views. He did not share most of the *specific* liberal political convictions, particularly if they affected him directly and would have demanded some decisive personal action.

CHAPTER 6

Family Life. The Move to Göttingen

The young couple's situation in Brunswick was not comfortable or secure, but there are indications that Johanna Gauss liked to live in the city where she had grown up and where she had friends and relatives.[1] Gauss and his wife belonged to the extended court of the Duke, in an old-fashioned and anachronistic way, like retainers at the court of an Italian Renaissance prince. Johanna was not idle; she had to manage the household, for Gauss does not seem to have been involved in its day-to-day affairs. Though we do not know the details, it is not difficult to imagine what life was like for them. There are many contemporary descriptions of similar attempts to protect one's private happiness and the satisfaction of a sheltered life by closing it off against the hostile intrusion of a turbulent world.[2]

In 1806, still in Brunswick, the oldest son was born and christened Joseph, after Piazzi, the discoverer of Ceres and indirect author of Gauss's fame. (The other children from the first marriage, Wilhelmine (Minna) and Louis, also owed their first names to astronomers: Olbers, who had first found Pallas, and Harding, who had discovered Juno.) To characterize Gauss's domestic life, we again quote from letters which were written in 1807 when Gauss visited Olbers in Bremen. First Johanna's letter:

I am sorry, my sweet darling, that my silence disturbed you, everything in our house went its normal course; I was very well, except that I wished you were here, Josepf (!) lacks nothing, he is very jolly, his new nurse, as I already told you, arrived Friday, she is a very straight, quiet person, an old spinster, but so very fond of children that Josepf was at ease with her from the first day, now he likes to be with her as much as to be with his mother, this I think proves that I can entrust him to her without reservation. Daily, he goes for walks and visits with her our relatives or his predecessors of whom there are quite a few, he likes this so much that he makes it very obvious by pointing at the door and by stamping if he does not want to stay indoors, his liveliness has much increased, everybody enjoys him and, according to Ebeling [the nurse] nobody believes that his delicate fine face is that of a boy, on the 26th arrived, without any ado, his 7th tooth, but in exchange the poor wretch lost his greatest good on Sunday, whenever I think of it, I am indescribably sad, but I was forced to decide quickly because it is uncertain how long Ebeling can stay ... my in-

decision about the date was the principal reason why I did not write to you on Friday, I also thought that letters for you would come, they came unusually late, after 9:30, the possibility that Harding could arrive earlier than you made me send the letter, excuse the confused envelope, I am in a great hurry.[*][3]

This letter crossed one from her husband in which he wrote:

I do not know, my dear Hannchen, how to apply more pleasantly the hours in which I am not particularly occupied, than by chatting with you, even if I do not have anything of importance to report. I shall continue to tell you how I am spending my time in Bremen . . .[4]

There is a postscript, answering Johanna's letter which had just arrived:

I very much enjoyed your dear letter of the 30th which I have just received. It is our invaluably good luck that our sweet Joseph can expect the critical moment in such good and reliable hands; when you receive this letter, the worst will presumably be over. Does he still give much attention to the study of the doctrine of equilibrium and motion?

The hardships of the journey did not affect my health, but the considerably changed diet (here, I eat a good four times as much as at home and people still complain about my poor appetite) initially caused me some obstructions which were removed by a laxative powder, now I am starting to get accustomed to the epicurean way of life. Olbers does not think that the pharmacy could do much against the weakness of my stomach, flatulence and obstructions, rather the cellar. Diet and the way one lives must help most against this type of bad health. He considers our ordinary red table wine to be unhealthy, and believes it could, even if it does not cause it, increase my occasional palpitation. For my stomach he recommends an occasional glass of Madeira, for a furthering of the opening a pipe daily at breakfast, otherwise walks, etc. Lukewarm baths would be good for me. The most salubrious effects, he thinks, would come from the occasional, repeated taking to the waters; who knows we might see each other again in Rehburg next year.—Olbers would be very interested to go with me on a trip to Paris; since both of us do not know how to appreciate the French theater and similar foolishness we could visit everything worth seeing in a few weeks and could complete the trip in approximately five weeks.

The news of the armistice between the French and the Russians seems to be confirmed and to point to an early peace, but in the meantime, the English have landed in Swedish Pomerania. What a mad time!

It is high time for me to dress: I have to finish in a hurry. Many regards to your good mother, and to all our friends as if I had named them one by one.[5]

Even at this early age—Gauss had just turned 30—he complained about his health and discussed his diet and digestion with his friends, particularly

---

[*] The reader has probably noticed that this letter is an answer to Gauss's letter of June 27, 1807, which was excerpted above.

Olbers. Later, we will encounter similar complaints for which he always had a sympathetic audience in the much older Olbers. Olbers was equally interested in discussions of his own health and his various ailments; though his friends were repeatedly worried and led to expect the worst, Olbers did not die until 1840, at the age of 81. Gauss seems to have been similarly robust; his most remarkable infirmities were temporary deafness in 1838 and an uncommon sensitivity to heat which seemed to cause him great discomfort. It was particularly bad during the geodetic work after 1818. Still, Gauss was a happy man. He was fond of his family and could look forward to many years of satisfying scientific work. Göttingen was a particularly good place to be, because scientific work, mathematics and astronomy, was more appreciated there than at any other university in Germany. The university's independence from clerical supervision and direct governmental interference, the timely emphasis on the natural sciences (there were, among the pre-Gaussian scientists, the mathematician Kästner, the physicists and astronomers Lichtenberg and Thobias Mayer sr, and in medicine, the poet Albrecht von Haller[6]) and its ample financial endowment gave it a leading place among the German universities.* Another, at the time unique, feature was the close connection between the Academy of Science (which we occasionally call the "Göttingen Royal Society"; its official name was *Königliche Societät der Wissenschaften*) and the university in Göttingen,[7] anticipating the romantic concept of the unity of teaching and research. In contrast, Prussia, and similarly Russia, followed the French example—its academies which had scientists like Euler, Lagrange, and Maupertuis for members were appended to the courts and did not interact with the educational institutions.

Science and education developed in Germany, unlike England, by reformation "from above" and not as the expression of the free will of the searching and emancipated citizen. This is, of course, yet another variation of one of the critical conflicts of 19th century Germany. For a long time, the battle lines in this conflict were not clear, and even the most conservative governments allowed the development of a liberal academic life at the presumably apolitical universities. This led to conflicts, and to unpleasant surprises for the governments concerned. It also, as is nearly superfluous to point out, contributed to the excellence of the German universities in the 19th and the early 20th centuries.

The political thrust of the Enlightenment was felt in Germany later and much less strongly than in France, and in a different way. The conflict only became obvious when the cry for constitutional government could no longer

---

* The other exeptional new university of the time was Halle (Saale), which played an important role in the internal development of Prussia, though it did not have Göttingen's privileges and was not free of clerical interference. The philosopher Wolff, influential and popular because of his popularization of Leibniz's ideas, was in fact dismissed for religious reasons. Even Kant, in Königsberg, 50 years later, had difficulties with the authorities.

be ignored after Napoleon had been defeated in the "wars of liberation". Even the most repressive of the Napoleonic governments in Germany had been forward-looking and liberal compared to many of the archconservative petty rulers who reestablished themselves after 1815.

The way things eventually turned out in France was actually not much different from the German example. During the Empire and after the Restoration the politically liberal forces lost many of their ties with the materially progressive powers which took over the government and then instituted a de facto conservative policy. The rise of positivism is quite typical of this development: the new philosophy released the forward-looking citizen from the obligation to be radical or revolutionary.[8] Germany anticipated this development implicitly: none of the German revolutionary efforts had much hope of success. In Germany, Kant's Janus-faced philosophy came in particularly handy: it allowed and even advocated radical, "Jacobinic" political ideas, but certainly no revolutionary actions. In its popularized and coarsened variety, it served as a kind of official ideology in the semi-absolutist German states throughout the 19th century and until 1918.

As a scientist, Gauss was very influential, and he was well aware of this; as a citizen, he was not interested in trying to exert any political influence— "political songs are repulsive songs" ("*politisch Lied ist garstig Lied*"[9]). This was a consequence of the objective political situation, but not a necessary one. This becomes clear by looking at the experiences and careers of some of Gauss's closest friends: Eschenburg, who had attended the Collegium Carolinum with him, became influential in the Prussian administration,[10] Olbers played an important role in the politics of his native Bremen, was a senator for a while and even a deputy for Bremen in the parliament in Paris, during the time when Bremen belonged to France; Gerling, a student of Gauss and later professor of physics at the university of Marburg, served for some years as a liberal deputy in the Hessian parliament in Cassel; finally, Lindenau, a dear friend, was for a number of years chief minister of the small principality of Saxony–Coburg–Gotha. Previously, he had been an astronomer, Zach's successor as director of the Seeberg observatory.*

The way Gauss understood politics reflected much of the spirit of the 18th century, and there are only a few areas in which he tried to be an independent and emancipated citizen. This is in stark contrast to his role as a scientist: As a mathematician and astronomer, he was the first man of his time, the "prince of mathematicians" even without the confirmation by his King. He started many of the new directions of mathematical research which were opened up at the beginning of the last century. This expansion of scientific research was not confined to mathematics: many of the diverse ways in which nature can be investigated only then developed into independent and acknowledged areas of research; we mention chemistry, geology,

---

* Lindenau is not a very good example—he was a baron, had served as an officer in the Napoleonic wars and was obviously "born to govern".

and geography. This development was an indirect consequence of the advances that had been made during the Enlightenment; at least in Germany, it was more directly connected with the so-called romantic movement.[11] Originally, this school of thought arose as a reaction to the "artificial" and "intellectual" spirit of the Enlightenment; one of its fathers was J. J. Rousseau. In discussing the concept of genius above, we have already described some aspects of the romantic age. The increased interest in nature and natural processes included the scientific side of the subject, but one should beware of identifying the study of science with a generally progressive attitude.* Politically and socially, the romantic age looked backwards, the virtues of the Middle Ages were discovered and Catholicism became fashionable again.† The movement was apolitical but it nevertheless left strong marks on the official politics of the continental powers. Their guiding idea of a "Holy Alliance" was a romantic (and quite unsuitable) attempt to establish a European security system based on the status quo and the divine rights of kings and governments. Gauss would never have called himself a member or product of the romantic school of thought—he despised its philosophy as far as he was familiar with it—but much of what he did and felt can well be seen as part of the romantic movement.

As a scientist, Gauss, insofar as his conscious reactions were concerned, subscribed to the convictions and creed of the Age of Enlightenment. One of his favorite books was J. P. Süssmilch's *Die göttliche Ordnung in den Veränderungen des menschlichen Geschlechts aus Geburt, Tod und Fortpflanzung desselben erwiesen*‡ which was published in 1741 and is probably the first German attempt to produce and interpret medical statistics.[12] Its preface, by the philosopher C. Wolff, begins with the characteristic sentences:

There is nothing more pleasant for man than the certainty of knowledge; whoever has once tasted of it is repelled by everything in which he perceives nothing but uncertainty. This is why the mathematicians who always deal with certain knowledge have been repelled by philosophy and other things, and have found nothing more pleasant than to spend their time with lines and letters.[13]

Süssmilch tried to prove the harmony of the creation (or at least of one aspect of it) by population tables, lists of births and deaths, etc. Gauss certainly followed him in this and would even have agreed to the well-intentioned introduction of the despised and wretched Wolff. This was the tradition that Gauss followed, though his view of mathematics was much wider: for Gauss, the whole of creation was like a book which was to be read,

---

* One of the leading romantic poets in Germany was Novalis, a mining engineer by education and profession.

† See, for example, the St. Mary's hymns by the poet Clemens Brentano after his conversion to Rome.

‡ *The Divine Order As Manifested in the Changes of Mankind by Birth, Death and Reproduction.*

and possibly understood, with the help of observation and mathematics. Mathematics is not a mere handmaiden of experience: she, the queen of the sciences, has two faces—as a basic tool for the understanding of nature (and not only of nature which centers around man) and as *jeux d'esprit*. But this disinterested penetration of nature as a serious, independent task was also a deeply romantic objective. Gauss never shared the humanistic and possibly shallow optimism of the Enlightenment. He was distrustful of mystical short-cuts (the trap of the "professional" romantic poets and philosophers) as well as classical idealism (leading to Goethe's antiscientific theory of color[14]).

CHAPTER 7

Death of Johanna and Second Marriage.
The First Years as Professor in Göttingen

In the fall of 1809, less than two years after moving to Göttingen, Johanna Gauss died of the effects of her last childbed, one month and one day after giving birth to her second boy. Poor Louis, as his father used to call him, followed his mother after a few months; shortly after his death, his father's engagement to Friderica Wilhelmine (Minna) Waldeck, the daughter of a law professor at the university, was announced. Gauss had been very happy in his first marriage; a year before Johanna died he described his domestic life in a letter to Bolyai:

... *The days go happily by in the uniform course of our domestic life: when the girl gets a new tooth or the boy learns some new words, then this is nearly as important as the discovery of a new star or of a new truth* ...[1]

Johanna's death was a terrible blow. Immediately afterwards, Gauss wrote to Olbers in Bremen:

You were kind enough to invite me for a visit after my wife was well again. She is well now. Yesterday evening, at 8 o'clock, I closed her angelic eyes in which I have found heaven for the last five years. Heaven give me the strength to bear this blow. Grant me a few weeks, dear Olbers, to gather new strength in the arms of your friendship—strength for a life which is only valuable because it belongs to my three small children. If the doctor permits it, I may follow this letter in a few days.[2]

Soon afterwards, Gauss set out for Bremen. The passages which follow are transcribed from a document in Gauss's handwriting, occasionally blurred by what might have been tears.[3]

Do you see my tears, beloved shadow? As long as I called you mine, you did not know any pain which was not mine, and you needed for your happiness nothing but to see me happy. Blessed days! What a poor fool I was to think that such happiness could last, to imagine you formerly incarnate and now again transfigured angel were destined to help me carry all the little burdens of life all life long? How did I deserve you? You did not need this earthly existence to become a better being. You only entered life to show us your light.

Alas, I was the happy one whose path the Impenetrable let be illuminated by your presence, by your love, by your most tender and pure love. Should I have ever thought you were one of our kind? Dear soul, you yourself did not know how unique you were. Placid like an angel, you endured my faults. O, if it be given to the blessed ones to be invisibly close to us who are but blind wanderers in the darkness of life—please do not desert me. Can your love ever cease? Can you withdraw yourself from this poor man whose highest good your love was? O you best, stay close to my spirit. Let your blessed tranquillity, which helped you endure the farewell from your beloved ones, impart itself to me, help me to be more and more worthy of you. Alas, what could replace you to the dear pledges of our love, what can replace your motherly care, your example, if you do not make me stronger and ennoble me so that I can live for them and not drown in my woe.

Oct. 25. Lonely, I am moving among the cheerful people who surround me here. If they make me forget my pain for a few moments then it will return later with double strength. I am out of place among your gay faces. I could be harsh towards you, and you would not deserve it. Even the serene sky makes me sad. Now, my dear, you would have left your bed, now you would walk leaning on my arm, holding the hand of our darling and enjoying your recovery and our happiness which we would read in each other's eyes, as in a mirror. We would dream of a beautiful future. An envious demon, no not an envious demon, the Impenetrable, did not will it. You in your bliss can already clearly see the mysterious purposes that are to be served by the destruction of my happiness. Is it not given to you to instill a few drops of consolation and resignation into the heart of the one who was left? While you lived you had plenty of both. You loved me so much. You wanted so much to stay with me. That I should not give myself over to my grief were almost your last words. Alas, how could I have avoided it? Please, do ask the Eternal—could he deny you everything—only for this: that your infinite goodness may always vividly stand before my mind so that, as well as I, a poor man, can, I may strive to follow you.[4]

On December 14, back again in Göttingen, Gauss thanked Olbers for his hospitality which he had enjoyed until the end of October.

Miss Waldeck had been a friend of Johanna Gauss, but we do not know whether this friendship meant more than the conventional relationship between a professor's daughter and the wife of a young colleague of her father. When Gauss proposed to her, she had just terminated another engagement for marriage, for reasons which we do not know.[5]

Gauss and Minna Waldeck married fairly quickly, but their engagement was not untroubled. The urge and wish to form a new union as soon as possible, to forget the tragedy of Johanna's death, to give the children a mother again, appear to have been more compelling than individual affection for his second bride. The role of the impetuous and longing suitor which he had to play again did not suit Gauss very well[6]—the letters which survive are quite cold and unemotional. We quote from the groom's first communication, a letter that was written on March 27, after he had already reconnoitered the terrain in a discussion with Miss Waldeck's mother:

I am writing this letter with a throbbing heart because my life's happiness depends on it. When you read it you will already know my wishes. How will you receive them? Will I not appear to you in a disadvantageous light since I am thinking of a new union, barely half a year after losing such a beloved wife? Will you think I am frivolous or even worse?

I hope you will not. How could I have the audacity to seek your heart if I would not fancy myself that you esteem me so highly that you would not ascribe to me any motive for which I would have to blush.

I honor you too much to wish to conceal that I can offer you only a divided heart from which the image of the glorious shadow will never disappear. But if you knew how the departed loved and esteemed you, you would understand that—in this important moment while I ask you to decide whether you can step into her place—I see the beloved departed vividly, acknowledging cheerfully my wishes and wishing me and our children happiness and bliss.

But, my dearest, I do not want to bribe you into the most important decision of your life. That a blessed spirit would look with deep pleasure on the fulfillment of my wishes; that your mother whom I made familiar with them (she herself will tell you why I did this) and your father who knows of them from your mother approve of my intentions and expect the happiness of all of us from them; that I to whom you were dear from the first moment I came to know you would be overhappy, all this I mention only in order to ask, to implore you not to take this into account, but to consult only your own happiness and your own heart. You deserve the purest happiness, and you must not be guided by any secondary considerations, of whatever nature, that are extraneous to my person. Allow me to confess quite openly that modest and easily satisfied as I am in my other demands on life, I do not know a medium way in my closest domestic arrangements. I must be either very happy or very unhappy: even the union with you would not make me happy if it would not make you completely happy.[7]

Minna Waldeck's social background was very different from Gauss's, and there were some complications before the marriage could take place. In his letter of April 15, Gauss prepares his future wife for a joint trip to Brunswick and tells her of his youth; he talks about his parents, the different jobs his father had and the relations of his parents to each other.[8] We quote from the postscript:

One more word: The reason why I did not write to my mother is that I wanted to surprise her; the reason why I did not want you to write is because my mother cannot read written things and you would certainly not want to bare your beautiful soul in front of people for whom it was not intended.[9]

The visit was followed by a temporary estrangement; the rift was healed by more letters to Miss Waldeck and her mother.

In August 1810, Gauss became the son-in-law of the professor and privy councillor Johann Peter Waldeck; the two surviving children from the first marriage had a mother again. Children followed quickly, in 1811 and 1813 the sons Eugen and Wilhelm, and in 1816 the daughter Theresa.

During his first years in Göttingen, Gauss received calls to the universities of Dorpat and Berlin. Dorpat was rejected because of the inclemencies of the Russian weather; the second offer followed Gauss's first contacts with the scientist, explorer, and politician Alexander von Humboldt,* one of the leaders in Prussia's revival after its defeat by Napoleon. Gauss did not accept a position in Berlin, but he did not turn it down either, and the negotiations and discussions about it dragged on for another 15 years. The matter will be taken up below in more detail.

A completely different experience was connected with a war tax levied by the French government in 1808. Gauss, who was subjected to it as a member of the university, was asked to pay ffrs 2000, a very considerable sum for a man who had just joined the university and not yet received his first salary. Without being asked, Lagrange in Paris and Olbers in Bremen offered their help, but Gauss did not want to accept any money from them. In the end, the contribution was paid by an anonymous donor who, somewhat surprisingly, turned out to be Count Dahlberg, formerly the arch-chancellor of the Roman Empire and then Lord Bishop of Frankfurt. There were other signs of his growing fame. In 1810, only two years later, Gauss won a medal from the Institut de France. He refused the money which accompanied it, but accepted the astronomical clock which was purchased for him by Sophie Germain.

More important than any distinction or reward was the way the Westphalian government strove to fulfill its promise to build a new observatory for Gauss. In 1810, it budgeted ffrs 200,000 over the next five years for its completion; in 1814, when the Kingdom of Westphalia ceased to exist, a lot of progress had been made although there was never as much money available as projected. Despite the troubled times, Gauss could think of acquiring the needed instruments. The first were bought in 1812 from the shop of G. von Reichenbach in Munich. In 1814, a heliometer manufactured by Fraunhofer followed (which Gauss never used) and in 1815 part of the inventory of a private observatory in Lilienthal near Bremen.[10] At the beginning of the winter semester of 1808/09, Dr. Schumacher from Hamburg came to study astronomy with Gauss. Schumacher was a lawyer, but his real interests were in astronomy. The following year, 1810, brought a number of

---

* The brothers Alexander and Wilhelm von Humboldt played a prominent role in the reorganization of the Prussian state, Alexander as a scientist and explorer, Wilhelm as a politician. They introduced radical reforms in the educational system of Prussia and specifically helped to establish the new university of Berlin. France was their great example; Alexander, at least a sympathizer of the French Revolution, had lived for many years in Paris and settled only late in life permanently in Berlin.

The old and the new observatory in Göttingen.

gifted students to Göttingen, among them Gerling, Nicolai, Möbius and Encke. Nicolai and Encke are known as astronomers, Möbius as a mathematician and an astronomer, Gerling as a physicist. Gerling actually developed, along with Schumacher, into one of Gauss's steadiest correspondents—Gerling, like Schumacher, invariably in the role of the pupil and listener, Gauss in that of the master.

It is often pointed out that Gauss was not interested in teaching, that he typified much more the research-oriented 18th century scholar than the educator and teacher.[11] Such a generalization is misleading, though Gauss himself liked to emphasize his disdain for teaching. There are many remarks to this effect in his letters, but it would be wrong to take them at face value and to apply modern standards*; teaching at Göttingen, the way Gauss experienced it, first as a student and then, during the first few decades of the 19th century, as a professor, cannot be compared with the notions of higher education which were developed in the second third of the last century. The majority of the students as Gauss knew them were uninterested, had little or no motivation, and even less knowledge.†

Though he may have disliked lecturing, Gauss seems to have been very willing to advise any actively interested student who approached him. There are many examples in the correspondence with Schumacher which show how Gauss did not mind explaining things in detail and repeatedly[12]; Gauss in turn expected from his students that they could work and think independently. Their own efforts, rather than lectures or explanations from their professors, had to be at the center of their studies. This is an attitude which is not hard to understand, but it is in conflict with the prevailing educational ideology of the late 19th century. This is one of the reasons why the conventional picture of Gauss as a teacher is distorted.

In addition to the explanations given in the correspondence, we can use Gauss's style in his published works as the clearest indication of his approach to presenting his work and to teaching. The style, after all, expresses the way in which Gauss wanted to talk to his readers and to explain to them his ideas. All modern mathematical literature is, in a sense, pedagogical, and Gauss is an influential proponent of this tendency. Still, his contemporaries and early successors found Gauss's style difficult and unmotivated. Even Felix Klein and Kronecker share this opinion. Accustomed as we are to the additional 75 years of increasing abstraction, we do not follow Klein; we cannot fail to recognize the author's pedagogical intent, expressed in the discussion of heuristic approaches, the many numerical examples, and, most importantly, his quest for the logically most direct (and not necessarily genetic) route in his arguments. The alleged difficulty of the style comes from the rigor (here used in the naive sense, not as a metamathematical category of reasoning) which Gauss applied in his papers and which was quite unique at the time; another factor is that motivations in a more general sense are missing—there are no great explanation or far-reaching surveys, offering surprising new views and revealing connections (though Gauss was quite fond of occasional cryptic hints at further ramifications of his theories

---

* Today's notions and attitudes towards teaching may be closer to Gauss's own feelings than the spirit which was prevalent between 1860 and 1960. It was in this period that Gauss's lack of interest in teaching was noted and criticized.

† During the first half of the last century, Göttingen started to lose its advantages over most other German universities. Educationally, Hanover fell back while other states, particularly Prussia, raised the level of higher education by improving their public school system. After a very short while this had a strong positive effect on the (also reformed) universities.

and concepts).[13] Gauss's motto *pauca sed matura—few but ripe*—is usually quoted in discussions of this kind. It may be more revealing to quote another explanation of his intentions as a writer (which were misunderstood even by his contemporaries). It is contained in a letter which Gauss wrote to Schumacher late in his life (February 5, 1850):

You are quite wrong if you believe that I only refer to the final polish of language and elegance of presentation. These cost comparatively little time, but what I mean is inner perfection. In some of my papers, there are such incidence points which cost me years of deliberation; nobody who sees them developed in a concentrated form would believe what difficulties had first to be overcome.[14]

This objective of Gauss, which he stated repeatedly, was often misunderstood. So was Sartorius' phrase (in *Gauss zum Gedächtnis*) that the scaffold had to be removed before the edifice of a mathematical work should see the light of the public.[15] Gauss had no desire to obscure his work; if he withheld a result, it was rather because he felt that it had not yet been completely understood—hardly a reprehensible reason for not publishing.

Section VII of *Disqu. Arithm.*

The object of this interchapter is to illustrate the preceding remarks about Gauss's style by means of a paragraph-by-paragraph summary of Sect. VII of *Disqu. Arithm.* Section VII is a coherent and nearly self-contained essay which was conceived as one of the earliest parts of *Disqu. Arithm.*, well before the algebraic apparatus of Sect. V was developed. It is 52 pages long in the format of the collected works and devoted to cyclotomy, the theory of the division of the circle; there are (for Gauss) unusually many remarks about further ramifications and consequences of the general theory.

§335, the introductory paragraph of Sect. VII, contains a general introduction, including a (famous and much noticed) remark about the division of higher *transcendental functions* (e.g., the lemniscate) which can be performed according to the same principles. It is part of Gauss's explicit program to simplify his concepts and methods as much as possible and to make his presentation short and clear.

§336. The object of this paragraph is to reduce the general case to the one in which the circle is to be divided into p parts, p a prime number. The general case follows from this special case if one solves certain polynomials.

§337. The trigonometric functions of angles of the form $2\pi k/p$, $k = 0, 1, 2, \ldots, p - 1$, can be expressed in terms of the roots of certain polynomials of degree p. Gauss lists the explicit equations for sine, cosine, and tangent, and shows why it suffices to consider the much simpler equation

$$x^p - 1 = 0$$

with the roots

$$\cos\frac{k}{p}2\pi + i\sin\frac{k}{p}2\pi = r.$$

It is in this paragraph that Gauss uses complex numbers for the first time in *Disqu. Arithm.* He does not define them explicitly or justify their use.

§338 quotes and explains the following lemma which is needed later. Let W be a polynomial with rational coefficients and roots $a, b, c, \ldots$. Let W'

be a polynomial with the roots $a^\lambda, b^\lambda, c^\lambda, \ldots$, λ an integer. Then the coefficients of W' are also rational. This is an important lemma which Gauss introduces here without motivation, presumably in order to avoid later interruptions.

§339. Let Ω be the set of the $p - 1$ imaginary roots of $x^p - 1 = 0$, p here and later always an odd prime. Let $r \in \Omega$. Then $r^\lambda = r^\mu$ if and only if $\lambda \equiv \mu \bmod p$. Let

$$X = x^{p-1} + x^{p-2} + \cdots + x + 1.$$

Then

$$X = (x - r^e)(x - r^{2e}) \cdots (x - r^{(p-1)e})$$

for any positive or negative integer e with $p \nmid e$. Other obvious but important consequences are

$$r^e + r^{2e} + r^{3e} + \cdots + r^{(p-1)e} = -1$$

and

$$r + r^e + r^{2e} + \cdots + r^{(p-1)e} = 0.$$

§340 contains a critical part of the proof, a fact that Gauss mentions explicitly. Let $\varphi(t, u, v, \ldots)$ be a symmetric function of $t, u, v, \ldots$. φ can then be written as a sum of expressions of the form $h^{(j)} t^\alpha u^\beta v^\gamma \ldots$. If one substitutes elements of Ω in φ, e.g., $t = a$, $u = b$, $v = c \ldots$, then φ can be written as

$$A + A'r + A''r^2 + \cdots + A^{(p-1)}r^{p-1},$$

where $A^{(i)}$ are uniquely determined and integers if the $h^{(j)}$ are integers. This relation can be generalized to

$$\varphi(a^\lambda, b^\lambda, c^\lambda, \ldots) = A + A'r^\lambda + \cdots + A^{(p-1)}r^{(p-1)\lambda}.$$

Moreover,

$$\varphi(a, b, c, \ldots) + \varphi(a^2, b^2, c^2, \ldots) + \cdots + \varphi(a^p, b^p, c^p, \ldots) = pA.$$

§341. Gauss shows in this paragraph that not all the coefficients of P and Q are integers if P and Q are two nontrivial, nonconstant factors of X. This implies, in modern language, that X is irreducible over the rationals. The proof is quite straightforward and follows from considering the roots of P and Q. The result of §338 is used in an essential way. There is an entry in Gauss's diary (#136) according to which he succeeded much later, in 1808, in directly proving the irreducibility of the equation $\Pi(x - \zeta)$, ζ ranges over the primitive roots of 1 for nonprime p.

§342 contains the program for the subsequent paragraphs. It consists of factoring X into polynomials of minimal degree. If $p - 1$ can be written as the product of $\alpha, \beta, \gamma, \ldots$, X can be split up into the corresponding factors. To simplify his formulas, Gauss introduces the abbreviation $[\lambda]$ for r^λ with $r \in \Omega$.

§343 contains the second essential step of the proof, namely the introduction of a primitive root g mod p. Let g be an arbitrary primitive root of n. Then the numbers $1, g, g^2, \ldots, g^{p-2}$ are congruent mod p (not necessarily in this order) to $1, 2, 3, \ldots, p-1$. This means that one can write Ω as

$$[1], [g], \ldots, [g^{p-2}]$$

and, more generally,

$$[\lambda], [\lambda g], \ldots, [\lambda g^{n-2}]$$

if $\lambda \neq 0$ mod n.

Let g' be another primitive root and $p - 1 = ef$. Let $g^e = h$, $g'^e = h'$. Then $1, h, h^2, \ldots, h^{f-1}$ are congruent (not necessarily in this order) to $1, h', h'^2, \ldots, h'^{f-1}$, or, more generally, $[\lambda], [\lambda h], \ldots, [\lambda h^{f-1}]$ are congruent mod n to $[\lambda], [\lambda h'], \ldots, [\lambda h'^{f-1}]$.
The sum $[\lambda] + [\lambda h] + \cdots + [\lambda h^{f-1}]$ is abbreviated by (f, λ); the set of roots in (f, λ) is called the *period* of (f, λ). The paragraph ends with the calculation of an example, the periods $(6, 1)$, $(6, 2)$, $(6, 3)$ for $p = 19$ and the primitive root 2. In modern language, §343 says that the Galois group of X is cyclic and generated by the automorphism $\zeta \to \zeta^g$.

§§344–351 are devoted to an investigation of the periods of the roots Ω.

§344. Two periods of the same length are identical if they have any root in common, Ω can be represented by the periods $(f, 1), (f, g), \ldots, (f, g^{e-1})$, $f = p - 1$.

§345. The product of two periods (f, μ) and (f, λ) of the same length consists of the "aggregate" (sum) of f periods, all of the same length. The paragraph contains additional statements about the products of periods, all of them direct consequences of their definition.

§346. Let μ, λ be numbers with $n \nmid \mu, \lambda$. Let p, q be periods of equal length. q can then be written

$$\alpha + \beta p + \gamma p^2 + \cdots + \vartheta p^{e-1},$$

$\alpha, \beta, \gamma \ldots$ all well defined and rational. Again, the proof follows from the definitions and the results of the preceding paragraph without much difficulty. §346 contains an explicit numerical example.

§347. Let F be a symmetric function of the type that was considered in §340. Let $t, u, v, \ldots$ be the f variables of F. If one substitutes the roots from (f, λ) in F for $t, u, v, \ldots$, F can be written as (see §340)

$$A + A'[1] + A''[2] + \cdots.$$

§348. More about the connection between periods and the polynomial which has its roots in a given period. By considering the periods and the polynomials which belong to them one obtains a decomposition of X in e factors of f "dimensions". The paragraph contains an explicit example:

Let $n = 19$. Let the sum of the roots in $(6, 1)$ be α. The sum of the products of any two roots is $3 + (6, 1) + (6, 4) = \beta$, of any three $2 + 2(6, 1) + (6, 2) = \gamma$, of any four $3 + (6, 1) + (6, 4) = \delta$, and of any five $(6, 1) = \varepsilon$. Since the product of all the roots equals 1, one obtains

$$Z = x^6 - \alpha x^5 + \beta x^4 - \gamma x^3 + \delta x^2 - \varepsilon x + 1 = 0.$$

§349. For large f, the methods of §348 which directly use the fact that the coefficients of a polynomial are symmetric functions of its roots are not very practical and quite cumbersome. Instead, one can use a theorem of Newton and calculate the coefficients from the sums of powers of the roots.

In §§350 and 351, Gauss continues his investigation of the connection between periods, roots, and polynomials. §351 contains two examples, one the problem of finding, for $p = 19$, the equation which has the periods $(6, 1)$, $(6, 2)$, and $(6, 4)$, and the other to find, again for $p = 19$, the equation with the periods $(2, 1)$, $(2, 7)$, and $(2, 8)$.

§352 contains a (verbal) summary of the intended further procedure. Not much more work is needed for a full investigation of Ω. First, decompose $n - 1$ into its prime factors:

$$p - 1 = \alpha \beta \gamma \ldots .$$

Now, let g be a primitive root of p, Ω can be divided into α periods, each containing $\beta \gamma \delta \ldots$ roots. One then determines the equations belonging to these periods and repeats this process until no further reductions are possible. After solving the reduced equations, if necessary with the help of tables, one reverses the process and eventually obtains the required angles.

§353. An example with $p = 19$ and primitive root 2.

§354. An example with $p = 17$ and primitive root 3. With the conclusion of §354, the main problem of the seventh section of *Disqu. Arithm.* has been settled. The last eleven paragraphs deal with interesting related topics and with applications.

§355. If a period contains an even number of roots, it is real; only the actual roots are imaginary.

§356 contains a first mention of the so-called Gauss sums

$$\sum \cos \frac{k\mathfrak{R}}{p} 2\pi - \sum \cos \frac{k\mathfrak{N}}{p} 2\pi = \pm\sqrt{p} \qquad (*)$$

and

$$\sum \sin \frac{k\mathfrak{R}}{p} 2\pi - \sum \sin \frac{k\mathfrak{N}}{p} 2\pi = 0 \qquad (**)$$

where $\mathfrak{R}$ runs over all the positive quadratic residues of p which are smaller than p and $\mathfrak{N}$ the corresponding nonresidues. When he wrote *Disqu. Arithm.*, Gauss was forced to keep the question of the sign of $(*)$ open. It was settled in a different paper, *Summ. Ser.* which was published in 1808 (cf. p. 30) Gauss sums occur in this context when one considers the quadratic equations that belong to the periods with $\frac{1}{2}(p - 1)$ roots. Gauss shows that these equations have the form

$$x^2 + x + \tfrac{1}{4}(1 - (-1)^{(p-1)/2}p) = 0;$$

they play an important role in two other proofs of the law of quadratic reciprocity (published only posthumously as part of *Anal. Res.*). In this paragraph, Gauss uses the polynomial to find the quadratic character of $-1/p$.

§357. Here, Gauss proves that the expression $(4x^p - 1)/(x - 1)$, p a prime number, can always be represented by $X^2 \pm pY^2$, X,Y rational integral functions of x. One obtains this result by studying the period $(m, 1)$ of the roots of

$$x^m - ax^{m-1} + bx^{m-2} - + \cdots = 0$$

and using the transformations of §348. The paragraph contains an explicit example; the result itself had already been announced in the fourth section (§124) of *Disqu. Arithm.*

§358 deals with the distribution of the elements of Ω over three periods (for $p = 3k + 1$). The calculation of the resulting polynomials is quite complicated but one obtains another interesting arithmetic result, namely the relation

$$4p = x^2 + 27y^2$$

for prime numbers of the form $6m + 1$. This was already known from the theory of binary quadratic forms.

§§359–360 contain the final step of Gauss's proof, the reduction of the auxiliary equations for the determination of Ω to ones whose solution involves only radicals ("pure" equations in Gauss's language; in general, the auxiliary equations are "mixed"). In modern language, this means that the Galois group is soluble by explicit decomposition. To do this, Gauss uses a well-known technique, the so-called Lagrange resolvent. Gauss's arguments are complicated and quite sketchy; contrary to what he promised, he never published a detailed presentation of this part of the theory. The posthumous paper "Disquisitionum circa aequationes puras ulterior evolutio" contains a few but not many more details. It breaks off after introducing Gauss sums which are, as we saw, treated in the separate paper *Summ. ser.* §359 contains the famous remark, added by Gauss in proof, that it would be futile to search for a general formula for the solution of equations of degree >4.

§§361–364 connect the investigation of the roots of X with the trigonometric functions of the angles which are the original subject of Sect. VII of *Disqu. Arithm.* The main problem is that of assigning angles to specific roots (without using trigonometric tables). The necessary procedure is explained in §361 in an elementary way; §362 deals with the determination of the other trigonometric functions from sine and cosine without using division. §§363 and 364 contain a short survey of the procedure if one deals with equations which have the trigonometric functions as roots directly instead of the much simpler equation X. The presentation itself is sketchy but §364 contains two detailed and informative examples, the cases $n = 17$, $f = 8$, and φ the cosine, and $n = 17$, $f = 8$, and φ the sine.

§365. This penultimate paragraph of *Disqu. Arithm.* finally gives the answer to the original question of the seventh section. The regular 17-gon and, more generally, the regular $(2^{2^\nu} + 1)$-gon can be constructed with straightedge and compass (*if* $2^{2^\nu} + 1$ is a prime). For $\cos(2\pi/17)$, Gauss quotes the following explicit expression.

$$-\tfrac{1}{16} + \tfrac{1}{16}\sqrt{17} + \tfrac{1}{16}\sqrt{34 - 2\sqrt{17}}$$
$$+ \tfrac{1}{8}\sqrt{17 + 3\sqrt{17} - \sqrt{34 - 2\sqrt{17}} - 2\sqrt{34 + 2\sqrt{17}}}.$$

He adds that it would be futile to try to divide the circle in the cases $n = 7$, 11, 13, 19,... but that the confines of *Disqu. Arithm.* would not permit him to prove this fact. None of Gauss's papers contains any indication of a complete proof of this last statement.

Above, we mentioned that Gauss knew very early that the 17-gon could be constructed geometrically. In 1801, before the publication of *Disqu. Arithm.*, Gauss sent a short manuscript to the St. Petersburg academy with an elementary proof of the constructibility of the 17-gon. This proof is, of course, not basically different from the substance of the seventh section but it avoids any of the advanced and abstract concepts which Gauss uses in the published version. It is quite likely that the St. Petersburg manuscript resembles Gauss's original proof. Its existence was not known to the editors of Gauss's collected works; a transcript with a short commentary (in Russian) was published in [Oshigova].

§366 contains a list of the 38 numbers <300 for which the regular n-gon can be constructed. They are 2, 3, 4, 5, 6, 8, 10, 12, 15, 16, 17, 20, 24, 30, 32, 34, 40, 48, 51, 60, 64, 68, 80, 85, 96, 102, 120, 128, 136, 160, 170, 192, 204, 240, 255, 256, 257, 272.

Gauss's Style

The second Brunswick period, from 1798 to 1807, was decisive for Gauss's further development. He learnt to assess his powers, to select areas of research to which to devote his time and energy, and to assert himself in the scientific world. His development was not confined to the theoretical or scientific sphere; when reaching out in this way, Gauss was faced with the question of what attitude he should choose when dealing with students or colleagues who sought his advice. Very soon it must have been clear to him that he could not expect to benefit substantially from such discussions and exchanges of opinion; almost always he was at the giving and not at the receiving end. It may have been a temptation, but Gauss never chose to close himself off and dissociate himself from ordinary scientific exchanges. He was very well aware of the obligation of the genius to explain (himself) and to present subjects which he himself was able to penetrate quickly in a way that was accessible to a wider audience. He accepted as a "partner" or colleague anyone in whom he could perceive honest effort and interest. The correspondence, particularly with Schumacher and Gerling, provides the best examples, but there other instances like the extraordinary posthumous essay "Zur Metaphysik der Mathematik".[1] Today, one would call it a didactic paper dealing with "foundational" questions like the definitions of addition and multiplication. It was probably written in the early 1800s, at a time when Gauss thought he would have to prepare himself for a career as a mathematics teacher. "Zur Metaphysik der Mathematik" is not very substantial and does not preempt the new foundational developments which occurred not much later and are associated with the name of Bolzano, among others. The essay reflects Gauss's mathematical research; it is conventional, and radical only in the sense that its reasoning is careful and that none of the traditional concepts are initially taken for granted.

Gauss's attitude towards his students (and others who sought his advice, like the Repsolds of Hamburg, the well-known builders of astronomical instruments) has been overshadowed by his reputation as a strict, even unfair critic of the work of others. His private judgments, particularly of colleagues, were often quite arbitrary and inconsistent; one perceives—*sit venia verbo*—the extravagance of the genius who cannot be sure what scales to use. In the correspondence, one finds acerbic remarks about Lagrange, Legendre, and Delambre who are reprimanded for their shallowness, platitudes, lack of understanding, and roundabout manner.[2]

We add here a few words about Gauss's style of writing. Earlier, his mathematical style was discussed; we now turn to linguistic and literary aspects. Most of his older published works are in Latin but there are some German papers; the posthumously published work is partly in Latin and partly in German. The most condensed and linguistically most difficult works are those he published in Latin; of them, *Disqu. Arithm.* is the most refined. To write in Latin was nearly anachronistic but the style corresponds well to Gauss's ideal of the most direct and pure line of thought. Gauss strictly avoids repetitions, redundancies, or anything that could be construed as rhetoric. The effects of the translation into an otherwise unloved language seem to have reinforced Gauss's tendency to write in a "scientific and classical" style. The celebrated Latin of *Disqu. Arithm.* was actually revised by Gauss's friend Meyerhoff, apparently because Gauss was not quite satisfied with his own efforts.

Similar statements can be made about the German papers Gauss published; they are slightly more accessible, having been spared translation into the lingua franca of science.* The German summaries of the Latin papers, most of them published in the journal of the Göttingen academy,[3] should be mentioned separately. These summaries do not contain detailed mathematical arguments but are rather informal surveys of Gauss's basic ideas, quite different from what one has been led to expect from him. The following quotation from *Summ. ser.* illustrates our point:

. . . Since, in the work quoted [Disqu. Arithm.], the investigation was quite far advanced and only the sign for an arbitrary value of k had to be determined, one could have thought that this was an easy step, after the main question was answered, and this the more so since induction leads to a very simple initial result: the roots in the formulas above are positive for k = 1 or for all values which are quadratic residues of n. Still, when one tries to prove this observation one encounters unexpected difficulties, and the process which leads to the determination of the absolute values of the series turns out to be quite unsuitable for the determination of the signs. The metaphysical reason for this phenomenon (to use an expression of the French geometers) is connected with the fact that the analysis of the division of the circle does not distinguish between the arcs $\omega, 2\omega, 3\omega, \ldots, (n-1)\omega$; all these arcs are treated the same way. This gives renewed interest to the investigation, and Professor Gauss was challenged to leave nothing untried to defeat this problem. After many and various futile attempts he succeeded in a way that was remarkable in itself . . .[4]

We will not discuss the style of the published nonmathematical papers (e.g., in astronomy or geodesy); they are clear and precise and in effect modelled on his mathematical research. They set the standards for the work of his contemporaries and succeeding generations of scientists in Germany.

---

* It appears that Gauss wrote the draft of some of his Latin papers in German; others were directly conceived in Latin and did not have to be translated.

The correspondence and the unfinished mathematical papers are freer, with more "motivation" and verbal explanation, but one should be careful not to overinterpret the difference. Even there, Gauss's arguments are usually concise and to an extraordinary degree correct; what is lacking may be final conceptual clarity and the typical simplification so often attained in his published works. Gauss used Latin extensively, even for many of his notes and the private diary, yet there is no doubt that he did not like it and that the translation made his style heavier and more complicated. If one takes these external influences into consideration one may well find that the different levels of style are nothing but the appropriate forms of expression in different situations.

Contrary to his statements about other mathematicians and scientists, Gauss's reviews of published works are usually mild and full of benevolent praise. The published reviews are decidedly negative only if the reviewed works are devoid of any mathematical sense or if experimental results were falsified so that they could not be used for further investigations. Gauss was quite an active reviewer; in the area of non-Euclidean geometry it was only by reviews that Gauss expressed his interest and convictions publicly.[5]

Comparatively many reviews are devoted to mathematical tables and to compilations of experimental results. Gauss's interest reflects the great need for an expansion of the numerical basis of mathematics, for tables of logarithms or number-theoretical tables of the type Gauss himself had attached to *Disqu. Arithm.* Gauss liked to discuss the usefulness of such books and even their formal organization and design. (See, for example, the review of Burckhardt's tables in Vol. II of *G.W.*, pp. 183 ff.) Apart from the fact that these tables were useful, Gauss seems to have enjoyed intelligently arranged collection of numbers, even if they were not of mathematical or scientific interest. Among his notes are lists of important historical dates, birthdays, biblical references, etc., which Gauss had compiled for his amusement.[6]

In Vol. IV of the collected works, one finds the famous reviews of the books by Schwab, Metternich and Müller (1816, 1822) in which Gauss intimated his conviction that non-Euclidean geometries existed. In these reviews, as in others, Gauss uses the opportunity constructively; occasionally, such a review seems to have been the starting point for a new and independent investigation. An example is the effect of Seeber's investigations[7] of the properties of positive ternary quadratic forms which Gauss reviewed in *Göttingische Gelehrte Anzeigen* in 1831. Seeber, a former student of Gauss, was inspired by the relevant passage in *Disqu. Arithm.*,[8] and Gauss used the opportunity to sketch again some of his ideas and to demonstrate a theorem rigorously which had been formulated but not proved by Seeber. Another effect of Seeber's book was a (short-lived) interest in crystalline structures. It showed itself in a brief presentation of the connection between ternary forms and crystalline structures which Gauss included in his review.

CHAPTER 8

The Astronomical Work. Elliptic Functions

In 1809, the bookseller Perthes of Hamburg published Gauss's *Theoria motus corporum coelestium in sectionibus conicis Solem ambientium.* The book, nearly 300 pages long in the format of the collected works, contains the sum of Gauss's work in theoretical astronomy but it does not always describe the actual methods which Gauss used in his own research. Like *Disqu. Arithm.*, *Th. mot.* was published in Latin; Gauss had written it in German but had to translate it because Perthes thought it would then sell better.* The subject matter of *Th. mot.* is the determination of the elliptic and hyperbolic orbits of planets and comets from a minimum of observations and without any superfluous or unfounded assumption. In his preface, Gauss mentioned the recent example of Ceres Ferdinandea whose discovery first caused the methods of *Th. mot.* to be developed.

Th. mot. is systematic to the point of being pedantic; it consists of two books, one with preliminary material and one with the solution of the general problem. The work is the first rigorous account of Gauss's methods for calculating the orbits of celestial bodies, directly deduced from Kepler's laws. Up to Gauss's time,[1] astronomers used ad hoc methods which varied from case to case, despite the fact that the theoretical foundations had been clear for more than 100 years. Gauss's essential contribution consisted in a combination of thorough theoretical knowledge, the unusual algebraic facility with which he handled the considerable complications which occur in a direct development of these equations, and his practical astronomical experience. Unlike his predecessors (including Newton, who had solved the problem by geometric approximation), Gauss did not presuppose which of the conic sections the orbit of the observed object describes. This complicates the computations but allows the treatment of new comets or planets without identifying them as such.[†]

---

* Probably for political reasons, Gauss avoided French, which would have been a natural first choice.

[†] Practically, it is very difficult to distinguish a planet from a comet on the basis of just a few observations.

Gauss on the terrace of the new observatory. Courtesy of *Städt. Museum, Göttingen.*

Before Gauss, Laplace had come closest to a solution of the problem, but his equations were so complicated that one could not hope to derive numerical results. Laplace's approach was straightforward, using directly Kepler's laws and the differential equations of the two-body problem. Gauss's derivation, which we sketch below, is very similar to Laplace's. His decisive new idea is to use the quotients between the areas of the *sectors* and the *triangles* which are determined by two radius vectors.

The four sections of Book I of *Th. mot.* discuss the relations between the different parameters which describe the movements of a celestial body around the Sun. Section 1 contains most of the necessary definitions, like radius vector, true and medium anomaly, eccentricity, and the trigonometric formulas which are needed to describe the position of a given body at a given point of its orbit. There are also practical hints on how to extrapolate numerical tables and how to approximate parabolas from ellipses and hyperbolas.

Section 2 is devoted to the determination of the geocentric locus of a celestial body as a function of three coordinates. Gauss starts out with the definitions of characteristic parameters like ecliptic and node: in §48, the seven elements are determined which characterize the motion of a celestial body. They are mean longitude ("epoch"), mean motion, the semiaxis major, eccentricity, longitude of the ascending node, and inclination of the orbit.

In the paragraphs which follow, Gauss derives and discusses the (trigono-metric) relations which hold between these elements. As in Section 1, there are criteria for the identification of the different conic sections. To conclude Section 2, Gauss established a differential equation for the motion of a ce-lestial body in geocentric coordinates and applies these theoretical consid-erations to a practical example. Again, there are remarks about the effects of errors in the observations.

The third section is devoted to the problem of computing the orbit from several observations (i.e., several points in space). The derivations are all elementary; in order to obtain usable results, expansions by power series are broken off after the first few terms (without discussions of the—obvious—convergence); the same happens with continued fractions. Special emphasis is given to the determination of orbits from two elements and to computing extensive examples. The introductory sentences of this section characterize Gauss's view of the scope of his work; they might be the motto of any of his papers or books:

> ... The discussion of the relations of two or more places of a heavenly body in its orbit as well as in space, furnishes an abundance of elegant propositions, such as might easily fill an entire volume. But our plan does not extend so far as to exhaust this fruitful subject, but chiefly so far as to supply abundant facilities for the solution of the great problem of the determination of unknown orbits from observations: wherefore, ne-glecting whatever might be too remote from our purpose, we will the more carefully develop everything that can in any manner conduce to it. . . . [Quoted from the English translation by C. H. Davis].

The fourth and last section of Book I deals with the case of several ob-servations all lying in a plane with the Sun. Again, trigonometric relations are derived; the model of a pyramid with the Sun at its apex is particularly helpful (§112). The section is short and concludes with the observation that the equations which were derived for this special case are useless for elliptical orbits.

After the preparatory work of Book I Gauss proceeds in Book II to attack the main problem of *Th. mot.*, the determination of the orbit of the celestial body from actual observations. The problem is solved in two steps, of which the first is to approximate from a minimum of three or four ob-servations and the other is to improve this first result with the help of further observational data. Sections 1 and 2 of Book II deal with the first task, Sections 3 and 4 with the improvement.

As mentioned above, seven elements of the motion have to be computed for the determination of the orbit. Gauss shows in Section 1 of Book II how to approximate six of these by using the absolute minimum of three obser-vations; the seventh, the mass, has to be determined independently. Each observation yields two independent parameters, longitude and latitude—this is the reason why three observations suffice unless the observed orbit lies in the ecliptic or is very close to it. To treat this case, the subject of the second section, four independent observations are necessary (because the three vanishing or nearly vanishing geocentric latitudes can no longer be used as independent parameters).

The first paragraphs of Book II,1 deal with preliminary problems like the effects of the actual position of the observer on the Earth or of the revolution of the Earth. The orbit cannot be determined directly because the ensuing equations are too complicated; to simplify them, Gauss decomposes the problem by considering the two equations X, Y in the unknowns x, y which in turn are functions of the elements of the orbit. X and Y are first formulated in a very general way and then refined by a converging iteration. At this point, Gauss reviews the relations which were obtained and explained in Book I and investigates how they can be used for substitutions and successive approximations. After gaining in this way a complete overview of all the existing possibilities, the precise equations for the elements are derived in §§140 and 141; they are complicated and lead to polynomials of the 8th degree. To simplify them, Gauss reviews the astronomical meaning of the different parameters and utilizes several conditions which had not been exploited so far. At this point, Gauss introduces the above-mentioned quotient between areas defined by sectors and triangles determined by two radius vectors. This quotient which characterizes the curvature of the orbit is suitable for iteration and yields the desired numerical results. Section II,1 takes up nearly a quarter of *Th. mot.*, so complicated are the necessary computations. Gauss's mathematical tools are no more advanced than algebra and (spherical) trigonometry. Section 1 ends with several examples; again, the margins of error (which, this time, are the consequence of the necessary simplifications) are discussed.

In the second section, Gauss considers the case of four independent observations, of which only two have to be complete. Methodologically, there is nothing new but we have already mentioned that this case is important if the orbit of the observed celestial body coincides, or nearly coincides, with the ecliptic of the Earth. In such a case, small errors in the observations can have enormous effects if one works with only three observations. Gauss again gives an example, this time with data for Vesta, a small planetoid with exceedingly small ecliptic.

The last two sections of the book are devoted to the task of refining the approximate orbits which are the result of the methods of the first two sections. Gauss publishes for the first time the method of least squares in Section 3—it is his most efficient and suitable tool for the improvement of the orbits. The method which is described (§186) as "the principle that the sum of the squares of the differences between the observed and the computed quantities must be a minimum" had been used by Gauss with great success in his calculations for Ceres. Priority of publication belongs to Legendre, but Gauss clearly made use of it before Legendre.[2] He gives two altogether new and original motivations in *Th. mot.* More about this controversial and for Gauss very important topic below (see pp. 138ff).

Section 3 is very short and disappointing to any reader expecting an exposition of Gauss's complicated and interesting approximation techniques. They were, at least in part, explained by Gauss in later papers; for a closer investigation, the incomplete computations of the Pallas perturbations have to be consulted.

In the equally short fourth section, Gauss makes some remarks about the perturbations of elliptic orbits which are caused by the major planets. Gauss does not go into any details, but he stresses the importance of the question for the precise calculation of the orbits and for the determination of the masses of the perturbing bodies.[3]

Th. mot. ends with several lengthy tables which clarify the relations between the different characteristic parameters of an orbit.

The major omission in *Th. mot.* is a lack of treatment of the parabolic orbits, a task that had earlier been solved by Olbers, and of Gauss's very efficient approximation methods. In spite of these deficiencies, *Th. mot.* became the most important and influential text of theoretical astronomy during the decades after its publication.

We mentioned that Gauss did not always follow the methods of *Th. mot.* in his own work but we will not go into the particulars of the historical development. Because *Th. mot.* was conceived as a systematic and definitive work, Gauss appears to have occasionally preferred the theoretically more direct to the indirect but practically more efficient route. A comparison with his "Summarische Übersicht . . ." (1809) is informative—one sees that his earlier approaches were much more heuristic; successful but formally less appealing.

This summary of *Th. mot.* concludes our discussion of Gauss's work in theoretical astronomy; his practical work will be discussed below.*

Before concluding our discussion of this period of Gauss's life we will sketch some hitherto undiscussed parts of his mathematical work, the roots

---

* We make here one last remark about the incomplete Pallas perturbations. Romantic biographers called Pallas Gauss's *Schicksalsstern*; this was the one major task which Gauss did not complete—he gave up after enormous calculations and after a great effort had not been rewarded. Apparently, his mistake had been to break off the expansions of the perturbations too early, after the third instead the fourth or fifth element.[4] This is quite ironic in the light of the fact that Gauss was otherwise prone to be too detailed in his calculations. Felix Klein reports[5] having seen, in the midst of his Pallas calculations, this remark in Gauss's handwriting: "Rather death than such a life" ("Lieber der Tod als ein solches Leben").

of which go back to the beginning of Gauss's mathematical life. Our remarks will not extend to the foundations of geometry and certain parts of applied mathematics; those areas will be summarized later in a different context.

Most of the work to which we refer here was summarized by L. Schlesinger, in his essay in Vol. X,2 of *G.W.*, under the title "Analysis". The notion of analysis is artificial in this context because Schlesinger uses the term to characterize several relatively disparate areas of Gauss's mathematical work. Following Schlesinger, we distinguish three main topics which appear to have inspired Gauss and with which most of his papers deal. They are the *arithmetico-geometric mean* (traditionally abbreviated to agM), the *hypergeometric function*, and the theory of *elliptic integrals*. The most important papers in this area are (* means "published posthumously"):

"Disquisitiones generales circa seriem infinitam ... Pars prior", 1812

"Determinatio seriei nostrae per aequationem differentialem secundi ordinis"*

"Methodus nova integralium valores per approximationem inveniendi", 1814

"Theoria interpolationis methodo nova tractata"*

"Determinatio attractionis, quam in punctum quodvis positionis datae exerceret planeta, si eius massa per totam orbitam ratione temporis, quo singulae partes describuntur, uniformiter esset dispertita", 1818

Several posthumously published smaller papers in Vols. III, VIII, and X of *G.W.**

It is obvious from this list that a large part of the work was published posthumously. This makes it difficult to follow Gauss's development in any detail and we shall confine ourselves to sketching his results, instead of reconstructing the line of thought which Gauss might have followed (as Schlesinger tried to do). By and large, we accept Schlesinger's conclusions.

Gauss appears to have had a very early interest in the arithmetico-geometric mean, but our knowledge of his development is based only on indirect evidence and occasional fragmentary handwritten remarks. Most probably, Gauss's original interest was numerical. Only later was it motivated by the results which one obtains after a few elementary calculations. The arithmetico-geometric mean (agM) is defined as follows: Let two numbers n,m be given. Then the arithmetic mean m' is defined by $m' = (m + n)/2$, the geometric mean n' by $n' = \sqrt{mn}$. By continuing the process of forming means, i.e., by forming $m'' = (m' + n')/2$ and $n'' = \sqrt{m'n'} \ldots$, one obtains two series of numbers with a common limit, $M(m, n)$. The expression "arithmetico-geometric mean" is Gauss's own and first occurs in a note which Schlesinger dates as probably earlier than 1797 and certainly not later than 1798.[6] This note deals with the derivation of the series

$$Tm(1 \pm x) = 1 + \tfrac{1}{2}x - \tfrac{1}{16}x^2 + \tfrac{1}{32}x^3 - \tfrac{21}{1024}x^4 + \tfrac{31}{2048}x^5 \pm \cdots,$$

which one obtains by forming the agM of 1 and $1 + x$. In the same note, Gauss proceeds to compute a sort of inverse function of $Tm(1 + x)$ and establishes the connection between the two series. Later notes contain further computations dealing with the connections between agM and elliptic integrals, a topic which will be discussed presently.

Lagrange, independently of Gauss (and vice versa), first introduced the agM function in the literature (1784/85). He used the so-called Landen transformation

$$y' = \frac{\sqrt{(1 \pm p^2 y^2)(1 \pm q^2 y^2)}}{1 \pm q^2 y^2}$$

as his starting point and applied it to the integrand of an elliptic integral, thereby obtaining an algorithm which led to an approximate computation of the integral

$$\int \frac{N \, dy}{(1 \pm p^2 y^2)(1 \pm q^2 y^2)}, \qquad N \text{ an arbitrary rational function of } y.$$

This is an expression which one encounters when studying the length of ellipses and hyperbolas. It is clear how elliptic (and also lemniscatic or higher) integrals and agM are connected: if one forms the agM of two suitable functions one is led to series of the type that occurs in the theory of the integration of elliptic integrals.

One of Gauss's first discoveries was the relation

$$\frac{1}{M(1 + x, 1 - x)} = \frac{1}{\pi} \int_0^\pi \frac{d\varphi}{\sqrt{1 - x^2 \cos^2 \varphi}}$$

which he used (with a new derivation) in "Determinatio attractionis" (1818) for the calculation of the secular perturbations, i.e., the nonperiodic part of the perturbations. This is the only place where Gauss published any of his results from the theory of the arithmetico-geometric mean.

Much of Gauss's fragmentary work was devoted to the theory of the lemniscate.[7] Among the most notable results were the representation of the lemniscate as the quotient of two entire functions P,Q, the explicit calculation of P and Q, and the correct identification of the two lemniscatic periods $2\tilde{\omega}$ and $2i\tilde{\omega}$. The first period is real, the second purely imaginary. P and Q are basically special cases of Jacobi's ϑ-functions.

sin lem, as Gauss called the lemniscate, was interesting for him, not only because it led to many intriguing functional relationships, but also because it provided a natural entry into the general theory of elliptic functions. One important clue was the relation

$$\tilde{\omega} = \int_0^1 \frac{dx}{\sqrt{1 - x^4}} = \frac{\pi}{M(1, \sqrt{2})} \qquad (*)$$

which Gauss accidentally discovered in 1799. This happened when he noticed that the values of the two expressions were the same.

The next step consisted of a generalization of $(*)$, $(**)$ yielding the period of the lemniscate in the special case $\mu = 1$.

$$\tilde{\omega} = \frac{\pi}{M(1, \sqrt{1 + \mu^2})} \quad \text{and} \quad \tilde{\omega}' = \frac{\pi}{\mu M\left(1, \sqrt{1 + \dfrac{1}{\mu^2}}\right)}. \qquad (**)$$

This formula is part of the theory of the inverse function (of the general elliptic integral). Gauss defined S by partial fractions, by analogy with a representation of sin lem which he had obtained earlier:

$$S(\tilde{\omega}\psi) = \frac{4\pi}{\tilde{\omega}\mu} \sum_{n=1}^{\infty} (-1)^{n-1} \frac{\sin(2n-1)\psi\pi}{e^{\left(\frac{2n-1}{2} \cdot \frac{\tilde{\omega}'}{\tilde{\omega}}\pi\right)} + e^{\left(-\frac{2n-1}{2} \cdot \frac{\tilde{\omega}'}{\tilde{\omega}}\pi\right)}}$$

He was also able to represent S as the quotient of two "ϑ-type" functions which he calculated explicitly.

Though Gauss used well-known techniques, his progress was considerable and led to the solution of one of the classical and most difficult problems of 18th century analysis. Gauss's exposition is strangely two-faced: there are no new techniques, all the methods use real variables, and there is not even a formal concern about questions of convergence. But many of Gauss's results are very deep and could fully be understood only much later, after the theory of elliptic functions was completed.

It is not quite accidental that Gauss published only a few of his results and in fact never did proceed to develop "general and complete treatment" ("die allgemeine und lückenlose Behandlung") which he intended to give according to his diary. In addition to external factors, like his growing involvement in astronomy, there was an inner reason why this program remained unfulfilled: the elliptic functions are multivalued. Although the assumption of a complex period, because of the connection between trigonometric and exponential functions, was certainly not too far-fetched for Gauss, questions of the range of the function proved to be insurmountable without, for instance, the theory of Riemann surfaces.

We have already seen that Gauss, during his investigations of the inverse functions of the elliptic integral, discovered the ϑ-functions (which were independently introduced and treated by Jacobi only a few years later) and explored their transformation properties.[8] Only sparingly did Gauss use integration in the complex plane though he was quite familiar with it and its geometric visualization.

None of these results about agM and elliptic integrals were published during Gauss's lifetime, and he did not challenge Abel's and Jacobi's priority. His private statements, as we know from the correspondence, were, of course, different. In one of his letters he claimed that a publication by Abel relieved him of the trouble of having to publish approximately one-third of his results in this area.[9] One of his most satisfying experiences was that the agM re-emerged in his approximation of the perturbations of Pallas—another example of how nature allowed the use of "pure," sophisticated mathematics for the decoding of its secrets.

Modular Forms. The Hypergeometric Function

There are connections, already seen by Gauss himself, between his analytic work and the theory of quadratic forms. Some fragments from the early 1800s show that Gauss was familiar with the rudiments of a theory which eventually found its completion in Klein's and Fricke's work on modular forms. Both F. Klein and R. Fricke were deeply involved in the publication of Gauss's work and had the opportunity to study Gauss's fragments closely.

For Gauss, the arithmetico-geometric mean probably provided the most direct and simple access to the theory of modular forms. The connection is established by the transformations which map the infinitely many "equivalent" forms of the agM into each other.*

Today, we cannot tell how Gauss proceeded—there are only unsystematic fragments most of which were published posthumously (with annotations by Klein and Fricke). They are not a firm basis for extrapolation, even if we could date them reliably. In this account, we follow the plausible reconstruction which Schlesinger gives in his essay. Gauss's starting point was in number theory; it appears he was led to the theory of modular functions from the reduction theory of quadratic forms. The following problem, from p. 386 in Vol. III of *G.W.*, is central. Let (a,b,c) and (A,B,C) be two equivalent forms with the negative discriminant $-p$. Consider now a function f with $f(t) = f(u)$ whenever $(t - u)/i$ is an integer or whenever $t \equiv 1/u$. This function, which he called *Summatorische Function*, was never explicitly defined by Gauss, but it is in fact the absolute invariant of the modular group of all linear substitutions $t' = (\alpha t - i\beta)/(\delta + i\gamma t)$, $\alpha\delta - \beta\gamma = 1$, $\alpha,\beta,\gamma,\delta$ integers.

Gauss's notebooks contain several illustrations which show that he was aware of the geometric side of the theory. The geometric representation of the modular functions is the result of the following construction. A specific form is identified with a grid in the complex domain by choosing the unit vector as one of the basis vectors. Then one considers functions f of the

---

* Obviously, a function is not uniquely defined as the agM of two other functions. There are infinitely many ways to represent a given function as agM of two other functions.

second basis vector that are identical on grids belonging to equivalent forms. f is a summatory function in Gauss's sense, if it is invariant under the actions of the modular group. The illustration below is derived from one of Gauss's prettiest sketches (cf. Vol. VIII, p. 103) with some corrections by Schlesinger. Inequalities [III] and [IV] describe the exterior of the two circles around $i/2$ and $-i/2$; [I] should actually read $-1 < y < 1$; [II] characterizes the exterior of the circle around $i/4$ with the radius $\frac{1}{4}$. If one sets $t = x + iy$, the figure characterizes the fundamental domain of a modular function $j(t)$. Arithmetically, the shaded area is the geometric locus of the points $t = (1 + ib)/a$ for the reduced form (a, b, c) with determinant -1. The function $f(t)$ (see above) assumes any complex value exactly once in this area. This result, already known to Gauss, was rediscovered by Dedekind. On pp. 102ff of his essay, Schlesinger gives a much more extensive explanation of Gauss's sketch.

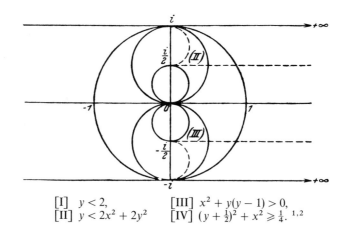

[I] $y < 2,$ [III] $x^2 + y(y-1) > 0,$
[II] $y < 2x^2 + 2y^2$ [IV] $(y + \frac{1}{2})^2 + x^2 \geq \frac{1}{4}.$ [1,2]

It is virtually impossible to give an adequate but concise summary of Gauss's work in the theory of modular functions. Because of conceptual questions, it will never be possible to ascertain what Gauss's notes and sketches actually meant and to what degree they preempted the later work of Dedekind, Fricke, Klein and others. We have not produced more than a sample of the concepts Gauss developed and the results he obtained; to go into more detail would mean a full reconstruction à la Schlesinger or Klein. It appears that this only posthumously published part of Gauss's work, with its many points of contact with the work of Jacobi, Eisenstein and others, was not without consequences. Dedekind, Klein, and Fricke knew it and seem to have been influenced by it. In spite of all his work, the different invariants which he calculated, the connections which he established between arithmetic, ϑ-functions, and elliptic integrals, Gauss does not seem to have conceived or to have seen the subject as a coherent theory. There is not much reason to assume Gauss would have been interested in such an abstract theory—the analogy to the group theoretical notions in *Disqu. Arithm.* comes to mind.[3]

Gauss's work on the hypergeometric function continued Euler's extensive investigations of its analytic properties.[4] Gauss's contribution is contained in two papers of which he only published the first. Probably written in 1811, it appeared in 1812 under the title "Disquisitiones generales circa seriem infinitam . . .". *Disqu. gen.* I starts out with the defining infinite series

$$F(\alpha, \beta, \gamma, x) = 1 + \frac{\alpha \cdot \beta}{1 \cdot \gamma} x + \frac{\alpha(\alpha + 1)\beta(\beta + 1)}{1 \cdot 2 \cdot \gamma \,(\gamma + 1)} x^2 + \cdots.$$

This may come as a surprise; Euler had already introduced the defining differential equation, which, one should assume, would have provided Gauss too with a natural way to introduce the function. He does not even mention it; instead, considerations about the convergence of the series immediately follow its definition. These considerations make use of the geometric series and show that Gauss did not confine his definition of F to real values of x. Next, Gauss derives the basic functional equation

$$\frac{dF(\alpha, \beta, \gamma, x)}{dx} = \frac{\alpha \cdot \beta}{\gamma} F(\alpha + 1, \beta + 1, \gamma + 1, x).$$

From this relation, a number of similar equations follow which concern the connection between F and the trigonometric functions (§§4 and 5). In §§7–11, the linear relations between the functions

$$F(\alpha, \beta, \gamma, x),$$
$$F(\alpha + \lambda, \beta + \mu, \gamma + \nu, x),$$
$$F(\alpha + \lambda', \beta + \mu', \gamma + \nu', x), \qquad \lambda, \lambda', \mu, \mu', \nu, \nu' = 0 \text{ or } \pm 1$$

are investigated; it is Gauss's obvious goal to state all the more important functional equations. §§12–14 contain investigations about the expansion of

$$\frac{F(\alpha, \beta + 1, \gamma + 1, x)}{F(\alpha, \beta, \gamma, x)}$$

using continued fractions. They are methodologically and substantially of no major interest.

§§15–18 are devoted to an investigation of the convergence of $F(\alpha, \beta, \gamma, 1)$, α, β, γ, real. Gauss then introduces the function $\Pi(x)$, a close "relative" of $\Gamma(x)$, characterized by the functional equation $\Pi(x + 1) = (x + 1)\Pi(x)$, and establishes its connection with F. Except for §3 and §§15–18, the paper is not too interesting theoretically, but very clear in its arguments and line of thought, two things which even today make it well worth studying.

The posthumously published second part was probably written immediately after Part I. It consists of 19 essentially finished sections. This time,

Gauss starts out with the defining differential equation

$$0 = \alpha\beta F - (\gamma - (\alpha + \beta + 1)x)\frac{dF}{dx} - (x - x^2)\frac{d^2F}{dx^2} \qquad (*)$$

and suitable boundary conditions. Gauss confines himself to arguments $|x| < 1$, because the function is no longer uniquely defined for $|x| > 1$; in the latter case, its value depends on the path on which one approaches a given value of x. The subject of the paper is the analysis of various interesting transformations and the values of the function at certain points. These investigations make it clear that Gauss was in full command of the techniques for integration in the complex plane, but he did not have the notions of analytic continuation and monodromy.

Schlesinger suggests that the papers on the hypergeometric function was conceived as the introduction to an intended comprehensive presentation of the theory of transcendental functions. The hypergeometric function played a central role in Gauss's thinking because he encountered so many special cases of the hypergeometric series in the theory of elliptic integrals and the agM.

We omit here a detailed discussion of Gauss's concept of the integral. Though this is obviously necessary for a real understanding of the two papers which were outlines above, we delay an explanation: this is better given within the framework of Gauss's papers in applied mathematics which we will treat below.

CHAPTER 9

Geodesy and Geometry

The years between 1818 and 1832 were dominated by the vast project of surveying the Kingdom of Hanover. Gauss himself directed the initial stages of this venture which took nearly 20 years to complete.

The contemporary interest in geodesy was essentially of a practical nature though it was also of a certain theoretical interest to determine by mensuration the true shape of the Earth. This question had already been taken up in the 18th century when extensive measurements had led to the universal acceptance of Newton's theory of gravitation.[1] In Gauss's time, additional quantitative results were still sought but there were also practical concerns. Geodetic work enjoyed official benevolence and liberal funding because the military and economic benefits of good maps were obvious.

The principal techniques for the various surveys were simple. Starting from a baseline of very precisely determined length, the area to be measured had to be covered by a grid of triangles whose vertices were visually connected. The actual work of surveying consisted of the establishment of such a grid and the precise determination of the angles. It is obvious that each "trigonometric point" had to be visible from a minimum of two directions. It was advantageous to exceed this minimum and to have, besides the regular rather small triangles, larger control triangles. This experimental work was time-consuming; even more so were, in the absence of any computing machines, the necessary calculations.

A certain part of Hanover had already been measured during the Napoleonic era and had been connected with the triangulation of the Netherlands. But the work had not been completed, its results were not sufficiently accurate, and many of the triangulation points were not even known any longer.[2] After 1815, all the major states of central Europe commissioned geodetic surveys. In the case of Hanover (and Gauss) the initiative came from Schumacher, who organized a similar survey for the Kingdom of Denmark. In 1818, Schumacher inquired whether Gauss would be interested in cooperation and the southern continuation of the Danish grid.[3]* Gauss,

---

* The Kingdom of Denmark included at that time the German provinces of Schleswig and Holstein; the latter had a common border with Hanover. Schumacher was in Danish services. He was affiliated with the University of Copenhagen and the observatory in Altona (near Hamburg).

who had performed some minor geodetic measurements during his second Brunswick period, was immediately attracted by the idea. He drew up a memorandum for his government, complete with a description of the project, the personnel needed, etc. A positive answer was secured quickly, and Gauss himself was made director of the project. The government gave the requested subsidies, and a few soldiers were detached to work as Gauss's assistants. At this point, Gauss certainly did not suspect that this operation would be the central task of the next ten years of his life, but the measurements were time-consuming and the ensuing difficulties much greater than anticipated.

The original plan foresaw only the connection of the Danish survey with the already existing results for Hanover, but that was soon dropped in favor of a completely original survey of Hanover, later to be expanded to include the territory of the independent city of Bremen. This last assignment presented its own difficulties because the coastal countryside was completely flat and practically on sea level.

The difficulties which had led to a premature end of the French survey stemmed from the peculiar topography of the kingdom. The country, especially in its western and coastal parts, is flat and covered* by large forests. It does not permit many of the essential long vistas, the establishment of trigonometric markers is difficult and in many directions impossible. The most problematic area is the so-called Lüneburger Heide, a sparsely populated stretch of land to the south of Hamburg, directly between Göttingen, where Gauss's base was, and the Danish triangles.

Gauss was not a merely nominal director of the project, he personally took charge of it. During the summer months of these years he rarely spent a night in his own bed and was often only a few nights in any one place, rushing from village to village, a victim of the inconveniences of the rural countryside (from which only two generations separated him) and of the heat of the summer. We see Gauss in his formal and proper attire, the indispensable velvet cap on his head, perspiring and directing his military assistants, bartering with farmers about the costs of the removal of a few trees which were supposed to obstruct the direct line of vision between two trigonometric points, organizing the distribution of instruments, etc. The evenings brought nearly daily correspondence with Schumacher (with complicated instructions where the mail should be sent and how it should be addressed), and endless computations. The measurements were made with the help of a small number of heliotropes, devices that had been invented by Gauss himself. Heliotropes were instruments with movable mirrors for reflecting the (dispersed) sunlight; with some minor improvements, they developed into a very efficient tool, allowing Gauss to connect much longer distances than his predecessors and to observe in less propitious weather, under cloudy skies and without direct sunlight. The correspondence with Schumacher, Olbers, and Bessel projects a vivid picture of the difficulties with which Gauss had to contend,

---

* Or, at least, *was* during the time of Gauss.

most notably in the Lüneburger Heide and the vicinity of Bremen. For a while, it was not even clear whether a suitable net of triangles was possible at all, and there were many weeks of anguish and (near-)desperation; happy were the moments when the fall of the last tree opened up the visual connection between two points, just as it had been half calculated and half guessed by Gauss. A quotation from a letter to Schumacher, dated August 30, 1822, illustrates the situation:

I have been looking forward to your visit from one day to the next and still hope that you will come because I will have to stay here for another 8 days; two directions are still to be established, the one to Wulfsode where Mr. Müller now is, and the one to Kalbsloh where he will go in a few days. One needs the latter because it is quite doubtful whether a line can be cut open between Hauselberg and Scharnhorst, since the very terrain of the Hassel may be too high. It is much more probable that things will work out for Kalbsloh, but I am reluctant to substitute Kalbsloh for Hauselberg because one cannot see Wulfsode from the former.

It is still not clear where I will go from here and I would have liked to discuss this with you first.

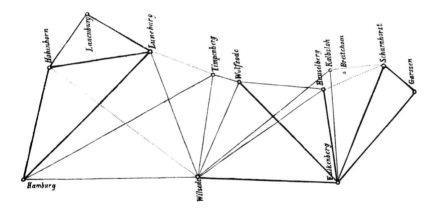

According to my preliminary calculations, Wilsede's elevation is m 12.3 above the level of the Göttingen observatory. In case you measured the zenith distance for the tower of St. Michael's church in Hamburg you may already, in a preliminary way, reduce everything to sea level. The distance Wilsede—Hamburg should be m 42454 ±.[4,5]

Two major theoretical works grew out of the Hanover survey, namely "Bestimmung des Breitenunterschieds zwischen den Sternwarten von Göttingen und Altona durch Beobachtungen am Ramsdenschen Zenithsektor" (1828) and "Untersuchungen über Gegenstände der Höheren Geodäsie" I and II (1843 and 1846). Both works exerted an enormous influence on the development of theoretical and experimental geodesy, but they are not of much interest except to the specialist. We confine ourselves to short summaries and start with the later work. Probably, it is the preliminary version of a projected, but never written major treatise on geodesy, à la *Th. mot.*

The method of least squares was Gauss's principal tool for the reduction of his geodetic observations. The theoretical fruit of his efforts, the theory of conformal mappings, is the subject of Gauss's Copenhagen Prize Essay of 1823, a paper which will be discussed below.

Part I of "Untersuchungen . . ." deals with the special case of the conformal mapping of an ellipsoid to a sphere; knowing this makes it possible to utilize ordinary spherical trigonometry for the purposes of geodesy. Small regions on the ellipsoid are mapped to the sphere by

$$f(v) = \alpha v - i \log k,$$

with i the imaginary unit and v an arbitrary point on the ellipsoid. The constants α and k have to be chosen in a way such that the distortion of the original area is minimized. The techniques which Gauss used were from complex analysis and spherical trigonometry. Two numerical examples, one with data from Hanover and the other from the Swiss survey, illustrate the general method. Part I of "Untersuchungen . . ." concludes with determining the azimuth at the other end, its geographical length, and the difference in longitude between the two points from the length of one side of a spherical triangle, the azimuth at one end and its geographical latitude. The paper is very slow-paced, obviously to make it accessible to someone without much mathematical or theoretical knowledge.

Part II of "Untersuchungen . . ." is devoted to the solution of this final problem of Part I for a triangle on the ellipsoid instead of the sphere. With the help of the mean values of the latitudes and azimuths and of tools from trigonometry and analysis Gauss obtains six formulas (§33) which solve the problem mechanically through tables which he had compiled himself and listed in an appendix to the paper. This technique of Gauss's enjoyed considerable popularity among geodesists and was in use until the end of the last century.

There exist a few very incoherent fragments for the envisaged major geodetic work, but they do not contain anything which would of interest to us here.

In *Bestimmung des Breitenunterschiedes* . . . Gauss determines the difference in geographical latitude between the astronomical observatories in Altona and Göttingen. Extremely careful measurements and a comparison of the respective zenith distances are used. The two observatories have practically the same geographical longitude; the determination of the difference in latitude supplemented Gauss's geodetic work and provided an additional control. The measurements were complicated, but the calculations were not; Gauss's skillful and systematic use of the method of least squares is noteworthy. The paper ends with a discussion of the irregularities of the

Earth's surface. In an appendix, Gauss determines its oblateness with the help of the method of least squares and based on the available data.

The development of the theory of conformal mappings is directly connected with Gauss's geodetic involvement. The other mathematically significant aspect of this period is the systematic use of the method of least squares. Both areas will be treated below in more detail.

There is a less direct connection to a resurgence and an expansion of Gauss's interests in the foundations of geometry and in differential geometry. The principal problem in the foundations of geometry was, at the time, the question of the character of Euclid's axiom of parallels as the only axiom in the Euclidean system that could not be understood with the help of a finite geometric construction. There had been many earlier attempts to clarify the axiom's role, the first already in antiquity. Most of them sought either to replace the undesirable axiom by an equivalent "finite" statement or to show that it could be derived from the other Euclidean axioms. Interest in the question increased sharply in the 18th century; in most of the works of this period, the authors tried to prove the dependence (and dispensability) of the axiom. Naturally, these efforts did not succeed, but they led to the discovery of various results which are actually theorems in non-Euclidean geometry. Among these propositions was the postulate of the existence of an absolute length, a curious fact which was proved in the second half of the 18th century by Lambert. Since this was supposed to be absurd the existence of an absolute length was seen as strong clue for the "correctness" and dependence of the axiom of parallels. Lambert, incidentally, did not think so—he had his own ideas and seems to have felt that a consistent system of axioms could exist without the critical axiom.

Kästner in Göttingen and Pfaff in Helmstedt were both interested in the problem, and Gauss may have had discussions with them as a student, possibly also with the astronomer Seyffer. There is some further information in the correspondence with Bolyai, and in statements which Bolyai made much later, after Gauss's death.[6]

Bolyai and Gauss both attempted to solve the question with the help of certain geometric constructions which avoided the axiom of parallels. Bolyai worked intensively in this area, even after his return to Transylvania, and in 1804 presented his findings in a letter to Gauss. His result was that the critical axiom was in fact not independent, but a consequence of the other Euclidean axioms. The central argument of Bolyai involved the following construction: On a given straight line, equidistant perpendiculars of the same length are constructed and their ends connected. In Euclidean geometry, one obviously obtains a line parallel to the original line, and Bolyai tried to show that if one did not make this conclusion one would be led to a contradiction. In his answer, Gauss praised his friend's work, but showed an essential flaw in the argument—Bolyai had, without justification, replaced an infinite construction which implied the axiom of parallels by a finite one. Gauss did not at the time indicate what his own beliefs were but he gave his friend no reason to suspect that he did not share Bolyai's conviction of the dependence of the critical axiom. Gauss wrote in his answer:

You want to know my sincere and frank opinion. And this is that your method does not yet satisfy me. I will try to make the critical point (which belongs to the same kind of obstacles which made my own efforts so futile) as clear as I can. I still hope that these cliffs will be navigated eventually, and this, before I die. For now, I am, however, extremely busy with other things ... [7]

So it is surprising to see that Gauss claimed in 1846 that he had been convinced of the existence of non-Euclidean geometries for the last fifty years (see p. 150). His first decisive and positive statements that we know of do not antedate 1816. In that year, Gauss discussed in a book review several spurious proofs which claimed to deduce the axiom of parallels from the other Euclidean axioms. Gauss was always very cautious in his public statements about any controversial question, and we can interpret the fact that he took a clear position as a certain sign that he had convinced himself that a non-Euclidean geometry could exist. The exact meaning of "had convinced himself" is unclear and we have, at least for the moment, to be vague.

We do not know how Gauss actually proceeded but he seems to have concerned himself with the development of what amounted to a consistent (trigonometric) model of hyperbolic geometry or, as he called it, transcendental geometry.* As far as we can see, Gauss was not interested in the philosophical question of the independence of the axiom of parallels; much more interesting was the actual geometric nature of physical space. Gauss's contemporaries did not properly distinguish between these two questions and were, like W. Bolyai, more interested in the philosophical side of the problem. The empirical question could clearly not be decided within Gauss's lifetime, a fact of which he himself seems to have been well aware. There are indications that Gauss "preferred" space to be non-Euclidean.[†8]

---

* This does not, of course, mean that Gauss had the modern notion of model, or, for that matter, of consistency. He derived theorems in analogy to the Euclidean case, and was satisfied that the new geometry appeared to be a coherent theory. Bolyai and Lobachevski, whose work will be mentioned below, did not proceed any differently.

The often-told story according to which Gauss wanted to decide the question by measuring a particularly large triangle is, as far as we know, a myth. The great triangle Hohenhagen—Inselberg–Brocken was a useful control for the smaller triangles which it contains. Gauss was certainly aware of the fact that the error of measurement was well within the possible deviation from 180° from which, under strict conditions, one could have derived the non-Euclidean nature of space.[9] It was, incidentally, Lobachevski who first proposed to investigate a stellar triangle for an experimental resolution of the question.

Another, more tangible, but at the same time even more speculative, aspect of this matter is the following. The shape of the Earth had been determined in the 18th century by very extensive geodetic measurements which had been organized to decide between Descartes's and Newton's theories of gravitation. Did Gauss see a possibility that these measurements would be pointless if our geometry was not Euclidean? A new fundamental discussion of Newton's theory was certainly not something Gauss would have looked forward to.

Such a radical interpretation of the old question was quite bold—we must recall that Kant, in his *Critique of Pure Reason*, had asserted that the Euclidean concept of space was an essential component of our mental framework.[10] Gauss occasionally mentioned Kant's position, but never, of course, accepted it. It is more remarkable that this difference of opinion did not affect Gauss's generally high regard for Kant's philosophy. This attitude was distinctly different from that of the (positivist or neopositivist) physicists of a later age who saw the edifice of Kant's idealism tumble down as a consequence of the theory of special relativity.

One of the reasons why Gauss seems to have preferred non-Euclidean geometry was the existence of an absolute length in non-Euclidean systems. In a letter to Gerling, he wrote: "One could use the edge of the equilateral triangle with angle 59°59'59", 9999 as unit length." ("... könnte man als Raumeinheit die Seite desjenigen gleichseitigen Dreiecks annehmen, dessen Winkel = 59°59'59", 9999.") This is well in line with Gauss's concern for absolute and independent units; other examples are to be found in the theory of magnetism (see below) and in the correspondence when he talks of the difference between right and left.[11] How satisfying it must have appeared to Gauss if Nature herself had provided an absolute unit for her most important dimension!

Gauss never published any original paper on the subject himself. In his correspondence, especially with Schumacher and Gerling, Gauss was quite outspoken, but at the same time concerned that his communications remained confidential. Among the most significant contributions was the work of F. A. Taurinus, a young lawyer who published two short monographs on the consequences of the deletion of the axiom of parallels. His work was unknown but Gauss, who had been made aware of it by Gerling, approved it, and seems to have studied it.

Several reasons prompted Gauss to hold back his convictions and not enter into a public discussion. His most important motive may have been that he did not want to get involved in what, in his eyes, was a completely irrelevant philosophical discussion of a question which he must have considered essentially undecidable.* To Gauss, the really interesting aspect was that physical space was suddenly not necessarily Euclidean. There is hardly any doubt that Gauss would have provided the theoretical framework if there had been at least a few experimental results or realistic schemes for observations. Many of Gauss's (often fragmentary) calculations published posthumously are concerned with "transcendental trigonometry"; they show

---

* It seems appropriate to give here a short explanation of Gauss's famous statement, in a letter to Bolyai, that he was silent because he was afraid of "the clamor of the Boeotians" ("das Geschrei der Böotier"). This should not be interpreted as an arrogant rejection of the opinions of the rest of mankind, because the phrase seems to be one which Gauss and his friends had used as students. The "Boeotians" are the uneducated, and the phrase reflects a disdain for unfounded arguments and discussions.

how interested Gauss was in understanding the consequences of deleting the axiom of parallels. His view of the problem was very different from ours; he likened geometry to mechanics, calling it, at its present stage of development, an experimental science and stressing the important role of intuition.[12] Despite his strong interest in the question, Gauss could not expect that there was a way to decide it experimentally. In his correspondence he expressed the conviction that more perfect beings than we humans would be able to see intuitively what the real geometry was—perhaps even we mortals could do so, after death.[13]

The first mathematically correct and quite "complete" collection of non-Euclidean geometric properties was published in 1831. Its author was Janos Bolyai, the son of Gauss's friend. Janos's paper was published as an appendix of the *Tentamen*, a textbook by his father who had sent it to Göttingen immediately after its publication.[14] Gauss's reaction is quite significant—he acknowledges Janos's mathematical achievement, his courage to publish something so controversial, but he does not mention at all the central question which one of the various potential geometries should be selected.

Gauss's last important statements on this topic were triggered by Lobachevski's publications (1841–1846). Gauss read some of Lobachevski's papers in Russian and others in German; he immediately recognized their importance and independence. In a letter to Schumacher he stressed their "genuinely geometric" character (see p. 150).

From 1815 on, Gauss's attitude throughout the discussion appears strangely rigid and indifferent; he seems to confine himself to variations on the theme that everything which others sent him had long been known to him. Janos Bolyai was particularly hurt when Gauss wrote to his father that he had been familiar with Janos's results for the last 30–35 years, a statement which was misleading if not false. It was in this context (letter to Bolyai of March 6, 1832) that Gauss used the ominous phrase that he must not praise Janos's work for *to praise it, would mean to praise himself*.[15]

Gauss's interest in non-Euclidean geometry was, as we saw, rekindled during the time of his geodetic work. There are connections, though not direct ones, between the geodetic survey and the foundations of geometry; similar connections exist with Gauss's work on differential geometry and comformal mapping, both of which were in fact inspired and essentially influenced by the survey. The two most important papers are "Allgemeine Auflösung der Aufgabe die Theile einer gegebenen Fläche so abzubilden, dass die Abbildung dem Abegebildeten in den kleinsten Theilen ähnlich wird" (*G.W.* VI, 1822) and "Disquisitiones generales circa superficies curvas" (*G.W.* IV, 1827). The first of these papers was submitted as prize essay to the Danish Academy of Science and will be cited as the "Copenhagen Prize Essay".*

---

* Following a request from Schumacher, the question had been formulated by Gauss himself, who refrained from participating in the contest for a couple of years. In the absence of serious competition he finally submitted his paper—which duly won the prize.

Analytic geometry, as we know, developed at the same time as differential and integral calculus. The first major contributions to differential geometry were by Euler; of his immediate successors, Legendre is the most important. Two of the original problems of differential geometry were the development of three-dimensional bodies, e.g., cylinders, in the plane and the general cartographic problem, i.e., to find the truest possible projection of the Earth onto a plane.

Gauss made substantial contributions towards the solution of both these questions, but his most lasting work in differential geometry consists of the identification and investigation of certain invariants that are intrinsic to geometrical objects, most importantly the measure of curvature. Gauss's ideas led to a completely new way of seeing the objects of differential geometry which was later called "intrinsic differential geometry". In this area, Gauss published his major results; the posthumously published material need not be discussed.

The original question which the Copenhagen Prize Essay sought to answer called for a derivation of all possible projections which could be used for the production of maps. The precise problem was to map a given, arbitrary area into another area in such a way that "the image is similar to the original in the smallest parts". Particular solutions were the stereographic mapping of the sphere which had been known since antiquity and Mercator's mapping. The problem had been solved in full generality by Lambert for the mapping of the sphere (globe) to the plane. In his essay, Gauss gives a complete solution, deriving a general conformality criterion for mappings between two arbitrary areas. His essential tools are integral transformations which are needed to reduce the square of a line element to a form familiar from the investigations of Euler and Lagrange.

Gauss's reasoning is direct and starts out with the following similarity condition (§4): ". . . that all infinitely small lines which start in one point of the first area and are entirely contained in it are proportional to the corresponding lines in the other area; and secondly, that the former have among themselves the same angles as the latter." Among other results, Gauss obtained a simple condition for the fact that two areas can be developed on each other. The Copenhagen Prize Essay ends with the discussion of three examples, the conformal mapping of planes into planes, the conformal mapping of a sphere into a plane, and the conformal mapping of an ellipsoid of revolution onto a sphere.

The analytic tools in the Copenhagen Prize Essay are the so-called Cauchy–Riemann equations,[16] and the already mentioned transformations of the squares of the line elements in the respective areas. It contains the first general treatment of conformal mappings, together with a number of examples, and the rudiments of the theory of isometric mappings.

Gauss's main work in differential geometry is "Disqu. generales..." (completed in 1827 and published in 1828). This is an important and

Gauss in 1828 (lithograph by S. Bendixen).

influential, but short, paper; despite the analogy in the title, it cannot be compared to *Disqu. Arithm.* and its place in the history of number theory. Gauss introduced several new concepts in *Disqu. gen.*, among them the already mentioned measure of curvature, and developed in §13 the basis for an important new part of differential geometry, intrinsic differential geometry. Gauss's two main sources of inspiration, as he himself wrote, were astronomical considerations, including spherical trigonometry, and theoretical geodesy. All his work is confined to three dimensions and to the Euclidean space E^3. His astronomical constructions led Gauss to the definition of the measure of curvature $K(A)$ in a point A of a given surface M in E^3:

$$\left|K(A)\right| = \lim_{\varepsilon \to 0} \frac{\text{area} \left(\zeta(D_\varepsilon)\right)}{\text{area} \left(D_\varepsilon\right)} \qquad [\text{``Gaussian curvature''}], \qquad (*)$$

with D_ε a compact ε-neighborhood of A in M, and $\zeta(D_\varepsilon)$ the corresponding hypersurface on S^2; it actually denotes the total curvature (*curvatura totalis* or *integra*) of the surface under consideration. This formulation is anachronistic, particularly (∗). Gauss's formula for the measure of the curvature is to be found in Sec. 10 of *Disqu. gen.*:

$$k = \frac{DD'' - D'^2}{(A^2 + B^2 + C^2)^2},$$

where A,B,C and D,D',D'' stand for certain differentials. Written out, this is, of course, much less clear than (∗). The subsequent problem consists in the explicit calculation of the measure of curvature; the most outstanding result is the famous equation

$$4(EG - F^2)k = E\left(\frac{dE}{dq}\cdot\frac{dG}{dq} - 2\frac{dF}{dp}\cdot\frac{dG}{dq} + \left(\frac{dG}{dp}\right)^2\right)$$

$$+ F\left(\frac{dE}{dp}\cdot\frac{dG}{dq} - \frac{dE}{dq}\cdot\frac{dG}{dp} - 2\frac{dE}{dq}\frac{dF}{dq} + 4\frac{dF}{dp}\frac{dF}{dq} - 2\frac{dF}{dp}\frac{dG}{dp}\right)$$

$$+ G\left(\frac{dE}{dp}\cdot\frac{dG}{dp} - 2\frac{dE}{dp}\frac{dF}{dq} + \left(\frac{dE}{dq}\right)^2\right)$$

$$- 2(EG - F^2)\left(\frac{d^2E}{dq^2} - 2\frac{d^2F}{dp\,dq} + \frac{d^2G}{dp^2}\right),$$

where E, F, and G are parameters which characterize the surface under investigation. The equation is called the "Gauss equation". It can be applied to any area M which is the image of an immersion $f: U \to E^3$, U an open subset of R^2. The geometric meaning of the equation is summarized in Gauss's famous *theorema egregium*:

If an area in E^3 can be developed (i.e., mapped isometrically) into another area of E^3, the values of the Gaussian curvatures are identical in corresponding points.

This theorem led to Gauss's famous and influential program for intrinsic differential geometry. We quote from Gauss's own summary (*G.W.* IV, pp. 344–345):

These theorems lead to a new way of seeing the theory of bent surfaces and open a wide, completely uncultivated field for investigation. If one does not interpret these areas as boundaries of solids, but as surfaces of one dimension less which can be bent, but not expanded, one sees that two types of essentially different relations have to be considered, namely those which assume a certain shape of the surface in space, and those which are independent of the shape of the surface. The latter are discussed here. According to what was discussed above, the measure of curvature is among them. One sees easily that figures on a surface, their angles, their volumes and total curvature, the shortest connection of points, etc., fall in the same category. All these investigations are based on the fact that the nature of a curved surface is given by an indeterminate line element of the form $\sqrt{(E\,dp^2 + 2F\,dp.dq + G\,dq^2)}$.[17]

Gauss's subsequent results deal with the properties of certain geodesics on a hypersurface in E^3, among them the well-known Gauss lemma which proves the existence of an orthogonal net of curves on M. In the summary, Gauss also notes the fact that the total curvature of a "geodetic triangle" is equal to the amount by which the sum of its angles exceeds 180°. This famous statement is now known as the Gauss–Bonnet theorem. *Disqu. gen.* ends with a generalization of a theorem of Legendre concerning the relation between the angles of a spherical triangle; this relation was also investigated by Gauss for the case of nonspherical triangles. A potentially important application concerned geodetic triangles; Gauss treated as an example his large "test triangle" (from his survey) Brocken/Hohenhagen/Inselberg for which he calculated the necessary corrections. Gauss himself was quite aware of the fact that this was only a theoretical exercise; on March 1, 1827, he wrote the following to Olbers:

In reality, this is quite unimportant because the unevenness of the distribution is negligible even for the largest triangles which one can measure on the Earth, but the honor of science demands that one understand the nature of this unevenness clearly . . . [18]

Strangely enough, he mentions this consideration only in a veiled form in the published paper.

Stylistically, *Disqu. gen.* is perhaps the most perfect among Gauss's shorter works; its approach is analytic, direct, and very concise. Gauss had every reason to consider it a well-rounded and reasonably complete presentation of his geometric ideas. The often lamented absence of a treatise on non-Euclidean geometry is not surprising in the light of the contents of *Disqu. gen.*: here, Gauss used all the sources which fed his geometric intuition—analysis, astronomy, spherical trigonometry and geodesy; the result was a series of concepts and theorems which reflected the full range of his geometric ideas and had a decisive influence on the further development of the subject.

Gauss's progress beyond the work of Euler, his most important predecessor in differential geometry, is substantial. Though Euler used the term "measure of curvature", in his work it is a global property, irrelevant if one wants, for example, to describe the curvature of the Earth. The motivation for Gauss's work in differential geometry is firmly embedded in his geodetic work, though it attained a generality far beyond its original scope.

We add here a survey of Gauss's work in an adjacent area of mathematics, the calculus of variations. This topic was first studied in the 18th century, when it was developed for the treatment of extremal problems in mathematical physics; for physical as well as for mathematical and philosophical reasons, the calculus of variations was a central theme of 18th century mathematics.

The tools for the solution of extremal problems are integration by parts and integral transformations. Two kinds of difficulties occur. First, the mathematical formulation of an extremal problem is often not clear, and it may not be obvious which of the several possible approaches will lead to a solution. It is particularly difficult to find the correct or most suitable boundary conditions. The second basic problem consists in developing, within the framework of the differential and integral calculus, an adequate and mathematically correct method for the formulation and execution of the variations. Historically, there were two fundamentally different approaches: variation of the domain of integration and variation of the independent variable. The first method is mathematically simpler, but of no use if one wants to treat a geometric problem. Lagrange found the main formula for the variation of the independent variable; the problem consists of the variation of

$$J = \iint V(x, y, z, p, q, \ldots) \, dx \, dy.$$

z is a function of the independent variables x, y; $p, q, \ldots$ are partial derivatives. Under some simplifying assumptions, Lagrange derived the formula

$$\delta J = \iint \Omega \omega \, dx \, dy + \iint \left(\frac{\partial(A + V\delta x)}{\partial x} + \frac{\partial(B + V\delta y)}{\partial y} \right) dx \, dy$$

with the notation

$$\delta \int Z = \int \delta Z, \qquad \delta \, dx = d \, \delta x, \qquad \delta \, d^2 x = d^2 \, \delta x,$$

(formal differentiation with regard to the independent variables $x, y, z, dx, dy, dz, \ldots$) and

$$\omega = \delta z - p \, \delta x - q \, \delta y, \quad \Omega = N - \frac{\partial P}{\partial x} - \frac{\partial Q}{\partial y}, \quad A = P\omega + \cdots, \quad B = Q\omega + \cdots$$

with

$$\frac{\partial V}{\partial z} = N, \qquad \frac{\partial V}{\partial p} = P, \qquad \frac{\partial V}{\partial q} = Q.$$

None of Gauss's papers is devoted directly and specifically to the calculus of variations; his most important efforts are rather contained in "Principia generalia theoriae figurae fluidorum in statu aequilibrii" (1829/1830) which one would consider a contribution to mathematical physics, and in "Disquisitiones generales circa superficies curvas" (1828), his basic paper in differential geometry, which has been discussed above. One sees in both these papers how familiar Gauss was with the theory and practice of integration and how well equipped he was to tackle a variational problem directly.

Mathematically, the problem of "Principia generalia . . ." can be reduced to the variation of the expression

$$W = \tfrac{1}{2} \iint [z^2 - g^2(x, y)] \, dx \, dy + (\alpha^2 - 2\beta^2) \iint \sqrt{1 + g_x^2 + g_y^2} \, dx \, dy$$
$$+ \alpha^2 \iint [z - g(x, y)] \, dx \, dy, \tag{$*$}$$

with α,β constant, $z = g(x, y)$ twice continuously differentiable, and of the expression

$$S' = \iint [z - g(x, y)] \, dx \, dy, \qquad (**)$$

where z and the domain of integration are to be varied.

Like his predecessors, most notably Euler, Lagrange used the δ-symbol freely and without defining it in a generally valid way. His results were correct, but he had to work with very narrow constraints in order to assure their validity. Poisson finally succeeded in 1816 in defining the δ-symbol in a consistent and general way (without leaving the known terrain of partial derivatives, substitutions, etc.), but Gauss's work, though later, seems to be independent of Poisson's results.

Physically, $(*)$ means the following. W is an expression which has to be at a minimum with respect to all the infinitesimal changes of shape of the constant volume of a liquid. $(**)$ is a formula for the volume of the liquid under consideration. The concrete setting is that of a homogeneous, incompressible liquid in a container whose equilibrium conditions are deduced from the principle of virtual displacements.

Though variational calculus was a much studied subject, Gauss gave a minimum of historical explanation; nor did he discuss the contemporary work of Ohm, who developed the details of the theory of the variation of the domain of integration.[19]

Gauss solved his problem in three steps:

(i) Formulation of the first variation.
(ii) Transformation of the first variation by integration by parts.
(iii) Derivation of a partial differential equation and of its boundary conditions.

The central question in the underlying physical problem concerns the optimization of the surface U, represented by the function $z(x, y)$ in Cartesian coordinates. Gauss's line of thought is most direct and essentially geometrical: U is varied by replacing each point (x, y, z) by another point in its vicinity. The introduction of the directional cosines ξ,η,ζ of the outer normal with regard to U yields

$$\delta U = \int dU \left((\eta^2 + \zeta^2) \frac{\partial \delta x}{\partial x} + \zeta\eta \frac{\partial \delta y}{\partial x} - \zeta\xi \frac{\partial \delta z}{\partial x} \right)$$
$$+ \int dU \left(-\xi\eta \frac{\partial \delta x}{\partial y} + (\zeta^2 + \xi^2) \frac{\partial \delta y}{\partial y} - \eta\zeta \frac{\partial \delta z}{\partial y} \right). \qquad (***)$$

The reason $(***)$ is so complicated is that Gauss did two things at the same time, combining the first variation with the calculation of the area of the hypersurface. In his derivation, Gauss did not separate these two different

---

* Here, as everywhere else in this paper, Gauss only considers the first variation.

and independent problems; one arrives, of course, at the same solution if one starts out from the formulas for the area of a curved hypersurface, which Gauss had proved much earlier, in 1813. Gauss used analogous direct ideas for the other first variations which are required. The calculations are clear and not difficult to understand.

The second step consists of manipulating (∗∗∗) using integration by parts. One obtains directly

$$\delta U = \iint \left(\frac{\partial A}{\partial x} + \frac{\partial B}{\partial y} \right) dx\, dy + \iint C\, dx\, dy, \qquad (\ast\ast\ast\ast)$$

with A,B,C homogeneous functions of x and y. With the help of purely geometric arguments, one can transform the first expression (∗∗∗∗) into a line integral along the boundary of the area U—this is, in fact, a generalization of Green's theorem and is known as Gauss's divergence theorem.[*20]

The third part of *Princ. gen.* is concerned with the resulting differential equations and their boundary conditions. It is not of interest here and will not be discussed.

Our review may appear to be quite unsatisfactory, (hopefully) for the same reasons why Gauss's work in this area had little historical impact. Contrary to the general development (and different from his own procedure in *Disqu. gen.*), Gauss did not separate general theory and specific problem; with this technical proficiency he appeared to relish the complications which arose, avoiding any appeal to the general theory that was known at the time.

This concludes our cursory account: Bolza, in his essay in Vol. X,2 of *G.W.*, gives much more information, most notably an additional analysis of *Disqu. gen.* which will not be further discussed here. The physical contents of *Princ. gen.* will be mentioned below, within the context of Gauss's work in physics.

---

* Gauss's explicit formula, which contains Green's theorem as a special case, is (∗∗∗∗∗)

$$\iint \left(\frac{\partial A}{\partial x} + \frac{\partial B}{\partial y} \right) dx\, dy = \int (AY - BX)\, dP$$

where P is the curve along the border of U. In this sense, Gauss's theory depends on the specific shape of U; also, A,B (and C in (∗∗∗∗)) are functions of U, but (∗∗∗∗∗) is identical with Green's theorem if U is a 2-dimensional plane.

CHAPTER 10

The Call to Berlin and Gauss's Social Role.
The End of the Second Marriage

Gauss's geodetic work did much to enhance his reputation outside the narrow circle of professional astronomers and mathematicians. But this is not the only reason why it would be wrong to consider this involvement a waste of time and a temporary aberration. Just as astronomers count him as one of their own, geodesists list Gauss as one of the greatest geodesists, a man who introduced new standards of observational efficiency and theoretical accuracy. Gauss's work set the standard for all subsequent efforts; his thoroughness, his tireless exertions in the face of major obstacles, his resourcefulness, particularly the invention of the heliotrope, were widely admired.

Schumacher's journal *Astronomische Zeitschrift** was an important outlet for Gauss during this period. With his astronomical and geodetic papers, Gauss was a fairly frequent contributor; he also advised Schumacher often and reviewed manuscripts for him. Indirectly, Gauss benefited by hearing through Schumacher of new developments, important gossip, and the activities of the other astronomers. Schumacher started his journal in a period of increased scientific exchange. In 1824, Gauss used the journal for a slightly different "social" purpose within the scientific community, when he issued a formal defence (*Ehrenerklärung*) of the Hungarian astronomer Pasquich.[2] Pasquich, the director of the Budapest observatory, had been accused of falsifying observational data. The accusation, originally by one of his assistants, had been confirmed by several prominent colleagues, presumably for personal reasons. Pasquich's moral and scientific reputation was in doubt, and it appeared that he would be ignominiously dismissed from his position. Gauss, whose opinion, if expressed, would be decisive, considered the matter carefully. When he published his conviction that Pasquich had been wronged, other astronomers joined him quickly, and Pasquich's reputation was restored with no further discussion. For Gauss, the step was extraordinary—

---

* Schumacher founded two publications, the *Zeitschrift* which appeared quite frequently, and the *Abhandlungen*. The latter was conceived as an annual progress report.[1]

he was not personally or scientifically involved, and Pasquich was only a mediocre scientist. But after discussing the affair with Olbers, Schumacher, and Bessel it did not take Gauss long to overcome his customary reluctance to make a public statement about a controversial and—strictly speaking— nonscientific matter.*

There was another round of serious negotiations with Berlin between 1822 and 1824/25. The moving forces on the Prussian side again were the Humboldt brothers, Alexander, the scientist and explorer, and Wilhelm, the enlightened politician and courtier.† Gauss's second wife Minna and her family had always looked to Berlin and encouraged him to move. The situation was particularly propitious at this time because the death of the permanent secretary[3] of the scientific section of the Academy made it possible to offer Gauss an adequate and well-paid position. Prussia was the leading and most vigorous of the German states, and it seemed only natural that the leading German mathematician and scientist should reside in Berlin. The negotiations were slow and involved, partly because of the tardiness of the Berlin bureaucracy, partly because Gauss was not directly participating. They were conducted on his behalf by von Lindenau, formerly director of the Seeberg observatory and now chief minister of a small principality in Saxony. Lindenau's task was not an enviable one because Gauss was never very explicit about his expectations or requests; in Berlin, on the other hand, decisions were very difficult to come by. Everything of importance, even (and especially) Gauss's salary, had to be referred to the King.[4] In a letter to Lindenau, written shortly before the negotiations came to an end, the Prussian general von Müffling defined the position which had been envisaged for Gauss:

We were all agreed that he should not be employed by the University. Because the minister had to have a title to justify the salary, he petitioned, with my support, the King that Gauss would be made an advisor to the minister concerning everything that pertained to mathematical education. Gauss would even be in charge of public projects like observatories, polytechnic institutes, etc. The King has agreed, and 600–700 Reichsthaler were allocated so there should be no more obstacles in this regard.

In addition, adequate travel and moving allowances will be granted.

Beside that of an academician, I think there is no position which would be more honorable; if Hofrath Gauss cooperates with the minister, he will have an enormous influence in the area of mathematical education. This is a vast field in which he can be very useful. The minister and the first councillors will have

---

* Pasquich did not benefit much from Gauss's magnanimity—he lost his position anyway. He was, however, allowed to retire honorably with a pension.

† This old-fashioned epithet is not out of place. The Prussian system, despite the innovations between 1810 and 1815, continued to be essentially feudal in the sense of the 17th and 18th centuries.

great confidence in him, everything else depends on himself. I have already
drawn up a plan for a polytechnic institute, and if this is in fact established, he
would have a major influence on its development. This would in turn provide an
opportunity to improve it even further . . .[5]

After the offer from Berlin with the final terms had been received, Gauss
confidentially informed the Hanoverian government to see whether it was
prepared to match it. At this point it appeared likely that Gauss would enter
Prussian service. But things developed differently. After some initial reluctance, Hanover increased his salary to what had been offered by Berlin; this,
together with assurances to further improve his situation and to improve
the observatory, decided the matter and Gauss did not leave Göttingen. The
outcome surprised and disappointed many of his "patriotic" friends, among
them Olbers, Bessel, and of course, Lindenau. For them, Berlin would have
been the natural place for Gauss. They wanted a progressive and attractive
Prussia which seemed to offer the only hope for the emergence of a unified
German state.

At the time, but even more so later, Gauss's motives were widely misunderstood. Kummer, not many years later, assumed that it was because of a few
Talers' difference that Gauss decided not to come to Berlin.[6] There is evidence
that Gauss would have gone if Berlin had been quicker and, more importantly, Hanover less responsive. Apart from this, there were clearly many
aspects of the Berlin appointment which must have made Gauss quite unhappy. Berlin was interested in his name and in his prestige, possibly in his
ability to stimulate and inspire younger colleagues. He was invited to be an
organizer and reformer who would raise the standards of higher education
to the level of the admired French schools. Yet it was clearly better for Gauss
to stay in Göttingen. There would be no upsetting move, with the attendant
interruptions of his scientific work, and he could expect more independence
and a better life if some details in his material position could be improved.
This is just what happened, but Gauss's attitude, already not clear to his
contemporaries, was to appear even less acceptable to the succeeding generations which had learnt to believe in the virtues of organized higher education
and progressive organizers.[7]

Yet another aspect of this affair is indicative of the general trend. The
Prussian bureaucracy was singularly inefficient in its dealings with Gauss;
Hanover, for all its reactionary conservatism, was much more flexible and
responsive. Gauss's demands had to be approved by the cabinet in London,
but a positive answer came forth immediately, without any of the circum-
stantiality and clumsiness of the Prussian officialdom. Once more, the old
system seemed to show its superiority, and the better organization and greater
efficiency for which Berlin stood did not appear to be real. In retrospect, it
is even surprising that Gauss should have been seriously interested in the
Berlin proposition, but such an offer could not be taken lightly; also, the
demands and concerns of his family may have been a factor.[8] As we shall see
later, Gauss was not completely immune to the temptations and tendencies

of his time, which demanded organizers and administrators. In Gauss's final reaction, his original instincts prevailed; socially and politically, Gauss was firmly rooted in the world that had existed during the first 20 years of his life.

Because of his reputation, Gauss was often asked for advice about vacant positions.[9] This was a typical demand for those (and later) times, and he took considerable interest in the merry-go-round of open positions and in the placement of his students. The close association with Schumacher kept him informed about the general scene, but his forcefulness and the respect which he enjoyed were certainly not the only reasons why his advice rarely went unheeded. Gauss's judgments were, as far as we know, conscientious and correct. He was concerned not only with a candidate's scientific qualification, but also with his abilities and skills as a teacher and organizer. This way, Gauss exerted considerable influence on the development of the German universities (and especially the affiliated observatories) during a most critical period. An enormous expansion of the system of higher education occurred at the same time, the general standards, especially of teaching, were raised, and the German system gained a leading role in Europe. Gauss's standards of rigor and accuracy were universally accepted by the middle of the century, and we can be certain that his direct influence contributed to this to a sizable degree.

Gauss's personal influence, together with the impact of his published work, helped to establish a firm base not only for the expansion of scientific activity but also for the technological and economic evolution of the middle and last third of the century. Though Gauss remained in small, isolated Göttingen, his real influence was no less than if he had gone to Berlin to start a new career in Prussia.

It was mainly in the private sphere that Gauss learnt to become "a child of his time". We have already described the circumstances which had led to his quick second marriage. Though it would be wrong to call it unhappy, we can certainly not compare it to the marriage with Johanna and even less with Gauss's recollections of it. After three childbirths between 1811 and 1816, Minna Gauss lost her good health and much of her capacity for housework and an active social life. Increasingly bedridden, she succumbed in 1831 to what probably was consumption. At the beginning of the marriage, Minna was quite independent: she was well educated, socially superior to her husband, and his only chance for regaining an inner equilibrium and the restoration of his domestic bliss. This last hope was never realized, be it because of Minna's long sickness, problems with the children, or, most likely, the sheer impossibility of recapturing the first happiness of youth. Minna's world was a new experience for Gauss, but he adapted to it quickly and completely. In 1811, when they married, Gauss was still the ingenious young scientist of humble origins, full of gratitude to the prince to whose munificence he seemed to owe everything that he was. The second marriage made him the son-in-law of a prestigious professor of jurisprudence, with an independently wealthy wife and an assured social standing.[10] In many ways, Gauss never changed, yet this second marriage left an indelible mark. Below, we shall come back to his financial transactions, mostly in government and private railway bonds, about which we read in the correspondence with Schumacher and Gerling.[11]

More conspicuous were Gauss's concerns about the careers of his sons, the awakening of what in German, is called *Familiensinn*. Gauss's efforts were only partially successful, and he experienced some bitter and humiliating defeats. The oldest son Joseph, apparently a capable engineer, was his father's assistant during part of the Hanover survey, and later supervised it himself for a while. He subsequently chose a career in the (Hanoverian) army, but his expectations were very limited, and even his father's repeated efforts could not bring about an overdue promotion.[12] The conservative orientation of the government blocked any hope for a further substantial advancement; Joseph resigned from the army and joined a private railroad company in Hanover. His contacts with his father were very sporadic, but the older Gauss enjoyed the distant veneration and the (somewhat limited) success of his son and repeatedly gave news of him in the correspondence. A curious detail illustrates the relation in later years: as a suitable present for his 75th birthday, Joseph gave his father an oil portrait of his only son, then $3\frac{3}{4}$ years old.[13] The relations with the other two sons were less serene and deviated from the fashionable ideal of family harmony.* Both Eugen and Wilhelm emigrated to North America after extended conflicts with their father. The alienation and separation were painful especially in the case of the older son Eugen, who had been forced by his father to study law, a career in which the gifted boy was not interested. He appears to have found it impossible to refuse his father directly; the conflict erupted because of Eugen's high life as a student in Göttingen and his gambling debts. There was a crisis in 1830, when Eugen became desperate and left Göttingen for an unknown destination. The last meeting between father and son took place in Bremen, where Eugen was found, but it was apparently without success. To both sides, emigration seemed the best solution, and Eugen embarked for Philadelphia.[†] This last conversation must have been very painful for the father—up to now, he seemed to have avoided direct confrontations and emotional involvement. He had repeatedly consulted his well-meaning but certainly not very perspicacious friend Gerling, relying more on the latter's advice than on his own instincts.[14] This correspondence about Gauss's problems with Eugen and

---

* The period between the Napoleonic wars and the revolution of 1848/49 is called *Biedermeier* in Germany. It is something like a precocious, German version of Victorianism.

† Our sources are quite incomplete because some letters which might have contained more information were destroyed or otherwise lost; Gauss is very cryptic in others. His attitude is not clear, but it appears he demanded either emigration or total and humiliating (the gambling debts were not to be honored) submission. Eugen preferred the former, emigration to the U.S.A., the appropriate haven for prodigal sons.

later Wilhelm leaves the reader with the impression of utter helplessness—
Gauss was neither able nor willing to understand his children's problems.
His own upbringing had been very different, but he was now a captive of the
social world into which he had moved, and of its values. Gauss had the objectives of the developing middle class for his family though they must have
been fundamentally alien to himself. It discouraged him that the younger sons
found it impossible to adjust to their father's expectations.

On his way back from Bremen, Gauss intended to see his son Joseph who
was then stationed with Hanoverian army in the small town of Peine near
Hanover. The conflict with Eugen had been aggravated by Minna's sickness,
which had recently taken a turn for the worse; from her sickbed she implored
her husband to go and visit her younger son Wilhelm.* Wilhelm was interested in farming—not a promising career in the eyes of his parents—and
worked at that time as an apprentice on one of the great Hanover estates.
Though he liked the work, Wilhelm was not very happy at any of the various
places where he was accepted, nor were his supervisors happy with him.
Wilhelm needed much attention, and the correspondence with Gerling tells
of various attempts to find new placements for the emotional and moody son.
There was no prospect of becoming independent and acquiring a good farm,
and Wilhelm finally resolved to try his luck in the new world. He emigrated
in 1832, after marrying a niece of Bessel, a connection which was quite surprising to his father and not very welcome.[15] While Eugen's farewell had been
stormy and left ill feelings on both sides, none of the possible conflicts came
out into the open this time. For both sons, emigration was supposed to be
and actually meant final separation—father and sons were not to see each
other any more, though there was some later correspondence across the
Atlantic, even with Eugen.[16] One of the reasons that made the conflict with
Eugen so bitter was that it took place during the last stage of Minna Gauss's
fatal illness; the marriage ended at a time when the dreams and hopes for
its children also foundered. Of Minna's children, the youngest daughter,
Therese, stayed with the father and kept house for him until his death in 1855;
only the two surviving children from the first marriage, Joseph and Minna,
had the orderly and predictable life which the parents hoped their other offspring would also be able to attain. Gauss's second marriage, as we saw, was
darkened by the conflicts with the children and, even more strongly, by the
long illness of his wife. There was certainly a psychological component to
her illness; though outwardly more independent, Minna never possessed
Johanna's self-assured and happy independence of mind. It is curious and
painful to see how an ever present insecurity pervaded Minna's life and forced
her into deeper and deeper despair. We quote from two letters which illustrate this development, one from 1811, the year of her marriage, when Gauss
visited Lindenau (letter to her husband of September 30, 1811), and the other

---

* We will quote from this letter below, see p. 116.

from 1830 when Gauss was in Bremen because of Eugen. First the earlier letter:

The children enjoyed your letter a lot, Joseph has asked perhaps ten times about his father, when does he come again? Minna, too, seems to be very concerned, she has specifically asked, will father have a gift for me?

Could I tell you, dear boy, how many sad moments I had while you were away, not counting Father's illness. Carl, my Carl, do you really love me? I feel it: my frequent indisposition must hurt you, but by God, I cannot overcome it;—also this oversensitivity. I cannot get over it, certainly, o certainly it is the consequence of an excessive sensitivity of the nerves, but it will, it must change, for, by God, it makes me most unhappy. Just have some patience, good boy, and do not withdraw your love, it will, it has to change, I do not want to go on living in such a melancholic mood.

How lucky that the end of the holidays and the presence of your mother direct you back to us, else, I am afraid, von Lindenau will give you so much to watch and to listen to that you would forget to return. Do not believe that I am jealous, I am so glad when you are satisfied, and I am sure that is what you are there. O God in Heaven—if I could only give you the happiness you expect, God knows, it is not lack of good will—but the strength, may heaven grant that the children develop into good men, then I would have fulfilled at least part of my destiny . . .[17]

Now to the second letter:

Though I had to make a firm promise to Himly not to write at all, I had to break it for you, good Carl.—How happy I am because your health is fine—alas, it is now my greatest good. My own health is in general much better, but you must not count on finding that I look much changed. Grief and illness have affected me so much that it will take some more time before my deep wrinkles disappear.— But it will come.—What you wrote about Eugen was such a consolation for me, God takes good care of us, I am so thankful to see that he let you find a ship, for we could really not expect that there would be one ready to depart.—Alas, this is the last thing you could do for him.—God help him. I feel it again, he is not dead for us, he is a prodigal son. I understand, my best Carl, that you had to stay longer, do not think that you have to write in advance the date of your return, this day will be a holiday and will bring light into the dark night which surrounds me. Now one more request, good, good Carl, do not deny it, arrange your return so that you can see Wilhelm, Mrs. Ihssen wrote, she requested it so urgently, both she and her husband are convinced that it would have a good effect on him, you also would enjoy seeing him, Carl, Carl, do it to for me as a favor, I suppose we both need some consolation. Joseph will not object, he is brought up well, honest and good, but Wilhelm has still to develop . . . He wrote to you, a letter full of the most sacred protestations how it would be his most fervent desire to please you and me. The Ihssens assure us they will take care of him even better now. Karl, could you refuse what I request? God, I have been so deeply humiliated, so deeply, I implore you do not deny what you can grant so easily.—God knows, I will do everything I can to rise from this deadly night, a night so full of grief. I cannot continue, God take you under his protection— Carl, Carl, do not refuse my prayers.[*][18]

For Gauss, the outlook at Minna's death was quite different from the situation he had to face when Johanna died, 22 years earlier. At that time, the end of his many dreams and hopes was shown by his helpless despair— now there was neither room nor reason for extreme feelings. Four days after Minna's death, Gauss wrote the following to Olbers.

The months after my last letter to you were a difficult time for my house. How long and how hard she had had to suffer before her heart could break. Finally, it did break. In the evening of the 12th, she departed from the woes of life, and today the earth received her mortal remains.

Both my daughters have been really helpful; my oldest son is right now continuing last year's observations, but I hope to see him in a few weeks here. My youngest son in Poppenhagen is starting to recover from a nearly fatal illness which he contracted six weeks ago.

I was consulted about Bohnenberger's old position and suggested Gerling, who in turn received an offer from Tübingen under very advantageous conditions...[19]

In 1809, when Johanna died, the future was much less predictable for Gauss than now, 22 years later. In 1809, Gauss had just started his career and could look forward to the fulfillment of many hopes. Only two years earlier, he had been installed as director of the Göttingen observatory; his fame was growing steadily; the erection of a new observatory was promised, to be designed and built according to his requirements and plans; the new social order, though not of his own preference, had brought him acclaim and all the scientific and social freedom he needed; his work and research were in a happy equilibrium between pure mathematics and applications. Still, the situation was transitory, and this in more ways than Gauss could expect—the observatory had not yet been built and should take much longer than anticipated; King Jerôme's regime in Westphalia was shaky and depended for its very existence on the further success of Napoleon's arms. It was not clear which way Gauss's scientific interests would develop and what the demands of society would be.

The outlook in 1831 was different. Now the future seemed to be determined and predictable. The changes which had come about had developed without much active participation by Gauss. Many things had happened during the 20 years of the second marriage, and Gauss had adapted to these events, seemingly more concerned about being left alone and pursuing what scientific work he happened to do (or for what work he happened to be asked) than

---

* The occasion for this letter is not entirely clear from the available sources, but it seems obvious that it was written when Gauss was in Bremen for his last interview with Eugen.

about actively determining his areas of research. Socially, the most significant development after Napoleon's defeat was the emergence of the bourgeoisie in Germany as a political and social force; in line with this development, the years of the second marriage mark Gauss's flirtation with the emerging middle class and an attempt to adopt its values and beliefs. The conflicts with his two younger sons are cases in point, as is the serious consideration which was given to the Berlin offer. It is difficult to say what would have happened if Gauss had accepted this position which, in its duties and in its spirit, was so unlike everything which Gauss was originally interested in. Perhaps it would not have changed him much; he would have fulfilled his tasks punctually and reliably, just as he dealt with the demands of the government in Hanover, sacrificing a good number of summers to the geodetic survey.* The survey, initially a welcome task, quickly deteriorated into an unending, repetitive chore. The letters from this period are full of complaints, ironically rarely about the labor of the survey itself, but rather the weather (the heat was particularly annoying), bad transportation or accommodation, and, quite often, bad health. Occasionally, a resigned complaint is made about Minna's ill health, one of the chief concerns and aggravations of these years.[21]

By conventional standards, this second marriage was good and could even be called happy, but it is impossible to tell to what degree Minna's afflictions alienated husband and wife. In the second of the two letters which we quoted above, she appears to betray a certain feeling of guilt about her illness; perhaps her husband blamed her for not to having fought it energetically enough.

This is not the place for psychological speculations, but it is unmistakably a mood of resignation and depression which prevails during the later years of the marriage and during this period of Gauss's life. This helps to explain the strangely pale picture which he projects during this time. Outwardly, it was a period of unceasing activity, filled with fruitful theoretical and practical work, with various trips and absences from home. In 1815, Gauss visited Munich with his son Joseph and an assistant in order to see some instrument-makers from whom he bought equipment for the new observatory. There were other trips, to Bremen, Berlin, and southern Germany. At home, a major change occurred when Gauss's aged and nearly blind mother moved to Göttingen to join her famous son and his family in 1817 (his father had died in 1808). Still, all these events take place as if behind a veil; the scene becomes no more vivid when when we know details, as we do in the conflict with Eugen. Even then, Gauss appears to be the passive spectator rather than an active participant; this is illustrated by his helplessness and his desire to resort to advice from friends. We quote from a letter to Gerling which was written

---

* Although by far his most voluminous, the survey was not Gauss's only official assignment. Later, an even less interesting task was connected with the standardizing of the Hanoverian weights and measures.[20]

on Nov. 13, 1831, a few months after Minna's death:

I have lost all my desire and will to live and do not know whether they will ever return. What depresses me so much is the relation to the good-for-nothing in (America) who has brought shame on my name. You know what message I received from him four months ago. I see it would have been good if I had followed your advice then: but I did not manage to answer him at all. Now a new epistle has arrived from him. It would be invaluable for me, dear friend, if you were at hand so that your proven friendship and your clear view could be of assistance to me. But fate did not wish to grant me this. Let me therefore, as well as possible, make use of your friendship from a distance because I can for obvious reasons not consult anybody who lives here. His letter is attached. I ask you, dearest Gerling, to tell me your opinion openly; in order to hear your unbiased judgment, I refrain from indicating what my impression is...[22]

His second wife's death, actually her slow death over more than a decade, the break with Eugen, and the loss of Wilhelm mark the ultimate defeat of any dreams Gauss may have nourished after Johanna's death in 1809.

It may not have been entirely accidental that the correspondence with Bessel came to a standstill and their friendship to an end very soon after Minna's death. We do not know exactly what went on but we know that Gauss was hurt when Bessel did not express his sympathies (which he never did, as a matter of principle). Bessel was a moody, difficult, and erratic man, and it would be wrong to blame only Gauss for the end of their friendship. But Gauss was increasingly sensitive, and he was now less than ever willing to acknowledge anybody as an equal partner in a scientific discussion. Of all the regular correspondents, Bessel and Olbers were the only ones who argued with him critically. The latter was an old man, disarmingly modest and pleasant. Gauss had much reason to be thankful to him, an obligation which he never found it difficult to remember. Bessel, however, owed much to Gauss, but he was a forceful man, and the correspondence with him was tiresome and dangerous.[23] In many respects, Bessel expressed the tendencies and the spirit of the 19th century; he had been instrumental in the attempts to bring Gauss to Berlin and to give him a role in the progressive scientific development of Prussia. In his desire to remove himself from perilous emotional experiences, Gauss gave up his friendship with Bessel; at the same time, he found it necessary virtually to break off relations with his son Eugen who appealed for his father's help (and money) shortly after his arrival in the U.S.A.*

Altogether, Gauss seems to have reduced his personal relationships to a bare minimum, embodying the stern father who was rigid and unapproachable for the good of his child. Again, Gauss relied on the advice and the good services of his friend Gerling. This dissociation from anything that might have

---

* Eugen enlisted in the U.S. army and asked his father for funds so that he could apply for an early discharge before his five years were up. Gauss did not give him the money.

disturbed him, his apparent equanimity, were not without their price, and other interests and concerns had to fill the gap.[24] Much of his energy was now given to introspection: the presence of his mother (who survived her second daughter-in-law by nearly eight years) helped him look back to his early life and early achievements. There is no direct proof for this change in attitude and outlook, but a number of clues exist, like his tendency to talk to admiring visitors about himself and to entertain them with childhood anecdotes.[25] Now we notice more and more often how Gauss cuts short the mathematical discussion of new results of others, claiming to have known these results all along but just not to have cared to publish them. There are statements of this kind in reference to the work of, among others, Abel, J. Bolyai, Eisenstein, and Jacobi.[26] It would be wrong not to mention that Gauss's claims were to a surprising degree confirmed by the posthumous publication of his unfinished works and the notebooks. Nevertheless, his sweeping comments are often off the mark—he remembers the extent of his own work incorrectly, gives wrong dates, or does not reflect the (mathematical, intrinsic, not external) reasons why he did not publish a certain result or a certain theory.[27]

By the end of the period which we are considering Gauss's increasing rigidity, the pervading mood of dissatisfaction, become more and more obvious, as did his desire to be left alone. We shall soon see that there were developments in another direction which helped him to regain some of his enthusiasm and creativity; here, we only wanted to show the reasons for his aloofness, the self-righteousness and the disregard for work of younger colleagues of which he is often accused. One factor in these reactions is Gauss's deep unhappiness, a loss of hope after which nothing appeared to be relevant any more or worthy of exceptional effort. Still, Gauss's life went on, and there is much left worth reporting.

CHAPTER 11

Physics

In 1831, Wilhelm Weber joined the faculty in Göttingen as a professor of physics. Gauss had first met Weber in 1828, at the annual convention of German scientists and physicians in Berlin. At that time, Weber was not yet a full professor; he taught as *Privat-Docent* at the University of Halle. Only 24 years old, he had impressed Gauss with his talk, and was now Gauss's first choice to succeed Tobias Mayer jr who had died the previous year.[1]

Scientifically and personally, Weber's coming suited Gauss well. He introduced Gauss to new areas of research, and induced him to involve himself in various physical, mostly experimental investigations. Their cooperation was very fruitful, and Weber's presence turned a period which otherwise might have been emotionally difficult and scientifically barren into an era of active research and rewarding discoveries.

Gauss had always been interested in physics, but most of his earlier investigations, apart from those directly connected with his work as an astronomer and a geodesist, had been quite theoretical.[2] His two most important earlier papers appeared in 1829, "Über ein neues allgemeines Grundgesetz der Mechanik" and "Principia generalia theoriae figurae fluidorum in statu aequilibrii". There are also, as we shall see, later important theoretical contributions, but then in connection with extensive and large-scale experiments. Gauss turned out to be quite interested in the organization of such large experiments and their intricate and costly setups, an attitude which was much more common in the 19th than in the 18th century.*

"Über ein allgemeines Grundgesetz der Mechanik" is a very theoretical work. It is only four pages long and was first published in a mathematical journal, the recently established *Crelle's Journal für reine und angewandte Mathematik*. Using ideas which go back to Maupertuis and d'Alembert,

---

* This is, of course, a superficial remark, and one should not read too much into it. In many areas, and certainly in those in which Gauss worked, the state of the art demanded costly experiments. Gauss was never extravagant or wasteful; he did not develop into a scientific organizer, but was indubitably fond of designing and overseeing costly experiments, even if their execution, as in the case of the geodetic survey, created enormous problems and demanded Gauss's constant attention.

Gauss derives a new, comprehensive extremal principle of mechanics, the principle of least constraint. The basic assumption of the paper is that any movement of a mechanical system, at an arbitrary moment, is in maximal agreement with the free movement; in other words, that minimal constraints always prevail. This approach allowed Gauss unified treatment of static and dynamic mechanical problems. The measure of the constraints of a system at a given time is the sum of the products of the squares of the dislocation in each point and its mass. All the concepts which Gauss used were carefully defined; he also established the connection between his principle and d'Alembert's principle of virtual dislocations which is closely related but weaker. One can see from Gauss's introduction how interested he was in extremal questions; he discusses in it the existing principles and gives an informal motivation for his own approach. The paper ends with a surprising remark about the analogy of his new principle to the method of least squares, i.e., the analogy between the laws of nature and the work of the calculating mathematician. The paper, as should be clear from the above, is of considerable interest to theoretical mechanics (see p. 140).

The second work from 1829, "Principia generalia . . .", is not as fundamental. It is a theoretical paper, not connected with any experiment. Gauss himself called it an exercise in theoretical physics; here, as in the other paper, one of Gauss's objectives appeared to be to show how much mathematics could contribute to the elucidation and explanation of nature. The subject matter of "Principia generalia . . ." is a discussion of the forces of attraction (molecular forces) which are effective over small distances and by which capillary phenomena can be explained. Gauss's predecessor here was Laplace, and Gauss introduces his work with the observation that Laplace's theory depends on two fundamental assumptions, namely a certain differential equation that describes the equilibrium of a fluid, and that, in equilibrium, the free surface of the fluid touches the wall of its container at a certain angle. The differential equation had been derived by Laplace, but his second condition could not be proved by him and had the character of a plausible heuristic assumption. It was Gauss's aim to derive Laplace's results in a rigorous manner and by a completely different route. His method is essentially an application of his principle of least constraint to virtual dislocations as they had been generalized by Fourier (from equations to inequalities). Mathematically, the paper is an application of the calculus of variations; physically, it proves that Gauss's principle is in fact stronger than d'Alembert's, a consequence of which Gauss himself was apparently not aware. We quote Gauss for one of his most interesting results directly, from the instructive summary which he prepared for *Göttingsche Gelehrte Anzeigen:*

Let s be the volume of the fluid, h the distance of its barycenter from an arbitrary horizontal plane, T the volume of that part of the surface of the fluid which touches the container, and U the volume of the other (free) part of this surface: then, in equilibrium, the expression

$$sh + (\alpha\alpha - 2\beta\beta)T + \alpha\alpha U$$

is at a minimum where α and β are certain constants which depend on the relation between the weight and the intensity of the molecular attraction of the particles of the the fluid towards each other and of the walls of the container towards the fluid.

We see that we obtain, as the result of a difficult and supple investigation, a condition for the equilibrium which is accessible even to the common mind and which shows the adjustments which take place in the conflict between the various prevalent forces. If gravity were the only force, the barycenter of the fluid would have to be as deep as possible, i.e., h would have to be at a minimum. If one were to set aside gravity and the attraction of the container and only consider the mutual attraction of the particles of the fluid, the surface would assume a spherical shape, i.e., $T + U$ would be at a minimum. If there were finally no gravity nor any mutual attraction among the particles of the fluid, it would spread over the surface of the container so that T would be at a maximum or $-T$ at a minimum. It is not surprising that a joint action of the three forces will minimize an aggregate of the three factors though one has to understand that a proper derivation of this theorem has to be based on complete and rigorous mathematical conclusions which largely depend on the nature of the molecular attraction.[3]

As an easy corollary one obtains formulas for the rise and fall of fluids in capillary containers. Gauss also investigated the effects of friction, less simply shaped containers, etc.

We have already mentioned that, viewed as a mathematical paper, *Princ. gen.* should be counted among Gauss's work in the calculus of variations. It also contains interesting results in potential theory. This was the context in which it was written; it is closely connected to Gauss's other work in this area.

Weber's coming marks the beginning of Gauss's systematic involvement in physics. He now became interested in experimental and practical questions which posed concrete technical and engineering problems. Still, his research retained its theoretical orientation, just as his astronomical and geodetic work had been accompanied by penetrating theoretical investigations. Naturally, Gauss's previous work in mathematical physics had many links with his new research; a common denominator is the fact that it can all be interpreted as application of potential theory to natural phenomena, principally geomagnetism and the theory of electricity. Gauss himself used the expression "potential theory"[4] and came, as we shall see, to recognize it, specifically Coulomb's law, as one of his basic tools for the mathematical and scientific understanding of nature, comparable to the method of least squares. There is a very suggestive analogy to mathematical astronomy, where the planetary movements, governed by Kepler's laws, are the prime example in natural philosophy for a (nearly precise) explanation of a nontrivial scientific phenomenon.

There are other reasons why potential theory became such a powerful tool in Gauss's hands. We have repeatedly pointed out how familiar Gauss was with the integration of complicated expressions, integral transforms, and similar techniques. Gauss's facility was particularly fruitful in potential theory, where he combined his mastery of these techniques with strong geometric intuition. He knew how to ask the correct and most direct questions and how to apply his analytic techniques. A detail from *Theoria attractionis* illustrates this point. In the course of his investigations, Gauss encounters two ellipsoids whose mutual attraction has to be determined.[5] Instead of trying to tackle the problem directly he shows that the initially arbitrary ellipsoids can be transformed into confocal ones without loss of generality. Gauss's solution is quicker and simpler than the work of his predecessors; moreover, it is the first strictly correct solution because Gauss is able to avoid certain divergent series which occur in the earlier proofs. Expansions into series are frequent and important in potential theory. So it does not come as a surprise that spherical functions, first introduced by Legendre, were Gauss's most useful tool. Gauss was interested in these and other special functions; among the posthumously published papers is a geometric interpretation of the spherical functions (cf. Vol. V of *G.W.*) which Gauss developed in connection with electrodynamic considerations. He remarks that the individual elements in the expansion of a spherical function can be interpreted as dipolic, quadrupolic, etc., contributions. The suggestive link between potential theory and complex analysis is not established explicitly, though Gauss was aware of a theorem which was equivalent to Cauchy's integral theorem. This should not be surprising, considering that Gauss was well aware of the various basic integral theorems in the real domain and also of the geometric representation of the complex plane. Prior to Gauss, it had not been clear at all that Coulomb's law was so basic, governing all kinds of different potentials. That it did was already much clearer to Gauss, but not completely, as his complicated electromagnetic formulas show (see below).

Gauss's work follows that of Laplace, he proved for the first time rigorously (in the "magnetic" paper "Allgemeine Lehrsätze . . ." (1840)) that

$$V = \begin{cases} 0 & \text{outside,} \\ -4\pi\rho & \text{inside a body with mass } M \text{ and density } \rho. \end{cases}$$

We can be quite sure that Gauss did not know Green's *An Essay on the Applications of Mathematical Analysis to the Theories of Electricity and Magnetism* (1828)[6]; he was familiar with Poisson's work but his proof was clearly the first correct one.

In 1832, Gauss started his investigations of the magnetism of the Earth. The investigation of terrestrial magnetism was one of the focal points of contemporary reseach; also, Gauss had earlier shown some interest without getting actively involved. The initiative seems to have come in this case not from Weber, but from Alexander von Humboldt, who tried to win Gauss's cooperation for his project to establish a grid of observation points all over the globe. Humboldt's was the first attempt at such a "global" experiment; he was the first to see clearly the need for a common standard of measuring

techniques, accuracy, and reliability. The results of these observations could
be expected to tell us about the distribution of terrestrial magnetism, local
and temporal changes in intensity, and declination and inclination; one
could finally expect to learn from these results the origins of terrestrial
magnetism and to be able to develop a satisfactory theory. Such a program
appealed to Gauss immediately.

Since his experimental work was essentially influenced by theoretical
considerations, we shall start with an explanation of the theoretical foundations of Gauss's work on geomagnetism. All three of Gauss's papers are
important; their titles are "Intensitas vis magneticae terrestris ad mensuram
absolutam revocata" (1832), "Allgemeine Theorie des Erdmagnetismus"
(1839), and "Allgemeine Lehrsätze in Beziehung auf die im verkehrten
Verhältnisse des Quadrats der Entfernung wirkenden Anziehungs- und
Abstossungskräfte"(1840). We also mention the monumental atlas of terrestrial magnetism, published by Gauss and Weber in 1840, with Gauss's
assistant C. B. Goldschmidt as coauthor.

It will suffice to confine ourselves to "Allgemeine Theorie des Erdmagnetismus" because a summary of this paper will provide us with an adequate
idea of Gauss's techniques. It starts with a discussion of the then current
theories of terrestrial magnetism, among them the assumption of the existence
of a single magnet in the center of the Earth or of two separate magnets
somewhere in the Earth's interior. Gauss does not enter into a thorough
discussion of these theories, but cuts it short and proceeds to define terrestrial
magnetism "empirically" as the force which pushes a magnetic needle into
a certain direction.

Let μ be the magnetic flow at a distance ρ from the source of the magnetic
force. The magnetic potential V is defined by

$$V = - \int \frac{d\mu}{d\rho}. \qquad (*)$$

Gauss derives several other expressions for V from $(*)$, again in integral
form, and gives an intuitive interpretation of the values which V assumes on
the surface of the Earth. In the next step, Gauss defines the magnetic poles
and shows that there can be only two of them. A discussion and the analytic
definition of the magnetic field lines follow. What has been summarized
above were not new results, but Gauss's work substantially clarified and
simplified the existing theories and their conceptual framework. Previously,
it had not even been clear why there should be only two magnetic poles,
and many of the then current theories assumed the existence of more than
two poles. Gauss's next theorem was new and of considerable importance
for the experimental scientist. It concerned the determination of the intensity
of the horizontal component of the magnetic force, together with the angle
of inclination; this characterizes the magnetic field. Gauss specifically shows
that one may calculate the western (sc. eastern) component of the magnetic
intensity from its northern (sc. southern) component if only the latter is

known for the surface of the whole Earth. The reverse is true with the additional condition that the northern (sc. southern) component is known for at least one connecting curve between north and south poles. Gauss obtained these (surprising) results by expressing the horizontal components as functions of the geographic longitude and latitude. The lack of symmetry is explained by the fact that the meridians run into each other in the north pole; there is no analogy for the latitudes. This result was not understood by his contemporaries; Humboldt, for example, for a long time considered Gauss's observations incomplete.[7]

Gauss calculated the vertical component of the magnetic potential V by expanding it into a power series with respect to the reciprocal radius of the Earth. One obtains spherical functions which can be evaluated with the help of the Laplace equation

$$0 = \frac{d^2V}{dx^2} + \frac{d^2V}{dy^2} + \frac{d^2V}{dz^2},$$

where x,y,z are rectangular coordinates of an arbitrary point. The basic assumption which makes the argument possible concerns the source of the magnetic force, which has to be located in the interior of the Earth. Then, Gauss's vertical component is a function of the distance from the surface of the Earth. Knowledge of the magnetic potential on the surface of the Earth suffices for the calculation of the potential at an arbitrary point; furthermore, as pointed out above, the potential is completely determined by its longitudinal horizontal component (sc. the latitudinal component and the longitudinal component in *one* point). This concludes the theoretical part of the paper; of the practical consequences, we mention only that Gauss was able to calculate, using his theory, the location of the magnetic south pole (which is, of course, close to the geographic north pole). In 1841, Captain Wilkes, the American explorer, reached it at a point very close to where Gauss had predicted it would be. "Allgemeine Theorie . . ." ends with extensive tables from observations and graphic visualizations of the magnetic field.

"Allgemeine Theorie . . ." is Gauss's central work on the theory of magnetism. His earlier paper, though of interest to the specialist studying Gauss's scientific development, is less clear and depends more directly on the work of others, most importantly on that of Poisson. Gauss's development was straightforward and without any surprises: in "Allgemeine Theorie . . ." the theory of magnetism appears as a mathematical theory, much more so than in the previous attempt, specifically as an application of potential theory. From what we explained above, it is obvious how Gauss made use of his superior mathematical versatility, once the basic definitions and facts were clear. "Intensitas vis magneticae terrestris . . ." contains Gauss's most original and perhaps most famous contribution to the theory of magnetism proper, the definition of an absolute measure for magnetic force. Gauss uses as a starting point that magnetism (the magnetic fluid, as he calls it) could (and

should) only be described by its effects; he proceeds to define the unit magnetic fluid as a force of such strength that it repels another unit magnetic
fluid at distance 1 with intensity 1. It is in this paper that the need for such
a definition is made clear for the first time in the literature; this need, though
obvious if not trivial in retrospect, was certainly not so obvious for Gauss's
contemporaries. They were too deeply involved in their efforts to fathom
the nature of the magnetic effect and to find a statement on which to base
a satisfactory theory.

For esthetic and practical reasons, Gauss was interested in obtaining an
efficient theory; his theoretical work was, as in astronomy and geodesy,
accompanied by extensive observations and measurements. This practical
work started before Weber's arrival and developed, after he came, into the
most intensive and fruitful genuine cooperation of Gauss's life.

Gauss's first observations were concerned with the local declination at
Göttingen and its changes with time. As was explained above, the horizontal
component $\mathfrak{H}$ of the magnetic intensity was what had to be determined.
The actual measurement proceeded indirectly, by observing the two effects
$\mathfrak{H}\mathfrak{M}$ and $\mathfrak{H}/\mathfrak{M}$, $\mathfrak{M}$ the magnetic momentum. Gauss chose this approach
because the physical dimensions of the magnetic needle do not enter into
the calculation explicitly.

The product $\mathfrak{H}\mathfrak{M}$ can be determined from the oscillation time τ of the
needle which is assumed to revolve freely around a vertical axis. τ can be
measured precisely; Gauss even takes the (usually negligible) effect of the
torsion of the thread from which the needle is suspended into account. For
the determination of $\mathfrak{H}/\mathfrak{M}$ an additional, auxiliary, needle is needed, again
rotating freely around a vertical axis. One measures $\mathfrak{H}/\mathfrak{M}$ by the deflection
which the auxiliary needle effects on the primary needle. This explains
Gauss's preference for heavy needles with long oscillation times—this actually was the only major detail in Gauss's technique dropped by the later
experimentalists.*

That terrestrial magnetism changed in intensity with time was a new, as
yet unexplained effect. The object of Gauss's research and that of his contemporaries was to map out the magnetic field of the Earth and to gather
information about the local, global, and temporal changes and perturbations.
This was why A. v. Humboldt initiated the establishment of a calendar for
the observations which determined the dates for the systematic measurement
of the magnetic declination by as many stations as possible. Gauss agreed
to cooperate with Humboldt's plan; soon after entering the field, he took
the initiative and changed Humboldt's design at certain critical points.

---

* There was an acrimonious controversy about this minor detail between the usually mild
Wilhelm Weber and the physicist J. v. Lamont. Gauss worked with needles of approximately
25 pounds, Lamont advocated the use of needles of a weight of 2 g or less. Schumacher, naturally
on the master's side, did everything to fan the debate and induced Weber's strong reply.[8]

Gauss's work led to the establishment of a magnetic observatory, the foundation of the *Magnetischer Verein* and its journal, and to the compilation and publication of the atlas of geomagnetism. These three "concrete" results of his involvement indicate why Gauss's work made Göttingen the acknowledged center of international research in this area shortly after Gauss had entered the field in earnest. The observatory, requested and designed by Gauss after he had seen Humboldt's observatory in Berlin, was erected in 1833. It was conveniently located next to the Astronomical Observatory; all iron fixtures, including the nails, were replaced by copper, which is nonmagnetic. In order to avoid drafts, all the windows and doors shut very tightly. With the new building, and his accustomed accuracy and skill in handling observational data, Gauss was immediately able to provide his colleagues with a wealth of reliable results. Soon after the observatory was completed, Gauss set out to improve upon Humboldt's procedures and techniques. Where Humboldt suggested measurements which lasted for 44 hours and had to be repeated every twenty minutes, Gauss introduced much briefer intervals of five minutes within a smaller total observation time. This change made the experiments much easier and the results more accurate; only now was it possible to identify many of the minor but interesting and important local perturbations. The outcome of Gauss's efforts to establish a worldwide net of observation points was the *Magnetischer Verein*. In 1837, Gauss and Weber started publishing their own journal, *Resultate aus den Beobachtungen des Magnetischen Vereins im Jahr . . .* , in which the relevant observations were recorded and published. These results constituted the basis for the geomagnetic atlas which Gauss and Weber compiled. The journal appeared six times, from 1836 to 1841.

Gauss was very active as correspondent and organizer during the magnetic measurement, even more so than during the geodetic survey. He constantly exchanged experiences with other observers, consulted about the optimal setup of the experiments, and discussed and corrected mistakes. Gauss was well aware of his superior knowledge and capacities; this attitude influenced his actions and is conspicuous in a letter to the venerable A. v. Humboldt which he wrote in 1833, when the magnetic observatory was still under construction:

I cannot really say that the insignificant experiments which I conducted 5 years ago when I stayed with you prompted me to occupy myself with magnetism. My interest goes back for more than forty years—it is as old as my interest in the exact sciences altogether but I have to confess that I devote my attention to a subject only if I have the means to penetrate deeply. This was not the case previously. The amiable relation which I have with our good Weber, his enormous kindness in putting the physics laboratory completely at my disposal and assisting me with all his wealth of practical ideas, merely made my first steps possible; in a sense, the first initiative came in fact from you, through a letter to Weber (at the end of 1831) in which you mentioned all the stations which were erected under your auspices for the observation of the daily variations.[9]

In the summary of the theoretical side of Gauss's geomagnetic work, one major theorem, or rather assumption, that Gauss made was omitted altogether. In *Intensitas* as well as in "Allgemeine Theorie . . ." one explicitly finds the statement that the distribution of a given mass over an area determines the potential of the mass in all points of the area in a unique way. Riemann called this hypothesis "Dirichlet's principle", a name under which it became known in the literature. Gauss, of course, was not concerned with any of the later hotly debated existence problems[10] and did not find it necessary to prove his assumption.

When Gauss became interested in magnetism, his research supplemented and continued his earlier geodetic work, contributing another important detail to the yet unfinished scientific description of the Earth. Additional motivation came from the important and fascinating contemporary work of Ørsted, Biot and Savart, Ampère, and Faraday. The results of their investigations laid the foundations for a new comprehensive theory, electromagnetism. When Gauss entered the field, there was a noticeable lack of a theory to explain all the newly discovered phenomena.

Gauss's work in theoretical electrodynamics is not well known, but it would be wrong not to discuss it. In the second part of his letter to Humboldt that we quoted above, Gauss reported about the most famous of his and Weber's experiments, the electromagnetic telegraph. Gauss and Weber actually developed two types of telegraph, one for which the voltaic chain provided the necessary potential difference, and another which worked with the help of a magnetic inductor. The latter was the more stable system, with clear and discernible reactions of the magnetic needle. Their first functioning telegraph, erected in 1838, connected the Astronomical Observatory with Weber's laboratory, spanning a distance of approximately 5000 ft. Gauss was quite aware of the practical potential of the machine, but never able to realize any of the large-scale experiments that he might have been interested in. His ideas were realistic and forward-looking—he suggested that railroad tracks should be accompanied by telegraph lines, possibly by using one of the rails as conductor (which technically would not have worked).*

Gauss did not penetrate deeply into the theory of electrodynamics—his attitude was that of an interested nonspecialist fascinated with this new branch of science. He discussed many of its phenomena and their theoretical explanations with Weber and made far-reaching remarks in his correspondence and in his notes, but did not develop a unified theory as he had done

---

* One should not forget that the idea of carrying signals over great distances was not alien to Gauss—during his geodetic survey, he worked extensively with acoustic and optical signals. Gauss and Weber's research on the telegraph was not original (as is often claimed). There had been earlier attempts and experiments, going back to the last decade of the 18th century, and even Gauss's idea of establishing telegraphic communication lines for the vast Russian empire had not been new. Friends called Gauss and Weber's experiments an aberration (*Irrweg*) and considered them frivolous and unscientific.[11]

so successfully for terrestrial magnetism. The correspondence, particularly with Olbers, shows how interested he was in the abundance of recently discovered effects; there is, for example, a discussion of the possibility of an electromagnetic motor, about which Olbers had read in newspaper articles and which Gauss considered quite improbable.*[12] There are also fragmentary investigations of the nature of the electromagnetic field in which Gauss made some contributions to a theory of electromagnetic "distance effect," a point of view to be taken up later by Weber and Carl Neumann. It was eventually discarded and superseded by Maxwell's theory. At the beginning of this century Schwarzschild showed that Gauss's line of thought actually presented a viable alternative which could be developed into a consistent and efficient theory.[13] At the basis of Gauss's consideration is a formula which describes, not quite correctly, the effect of two electric charges on each other in the form of a partial differential equation. The argument is very sketchy, and the notes were never meant for publication, but we can see how Gauss tried to apply his knowledge of potential theory to electromagnetic phenomena, without involving himself in the physical details.[14]

---

* These reports concerned experiments which were supposed to have taken place in North America. Gauss was incredulous; though he agreed that it was theoretically conceivable that such a motor could be constructed, he did not believe that it could be stronger than a mouse.

Gauss's Personal Interests After His Second Wife's Death

Predictably, the death of his second wife did not interrupt Gauss's involvement in the painstaking details of experimental research. Practical activity of this kind was a welcome distraction; he devoted to it the major part of his capacity for concentrated scientific work. Despite his involvement, Gauss was far from one-sided. Weber's sister, who was in charge of her bachelor brother's household, tells us that the great man had learnt how to move in society, how to be polite and to be a gentleman. He could talk about many things and insisted that no scientific problems be discussed in her presence. "So much," she remarked, "was he a man of the world."[1]

The distant observer of today finds it difficult to discern where Gauss's real interests were and whether his reactions, as we know them from the correspondence or from the reports of visitors, were just the automatic response to some impulse from the outside or the expression of genuine concern. There are only a few situations in which one can be certain of an emotional involvement—it appears that Gauss tried to avoid this whenever possible. We already know two conflicts in which Gauss was highly irritable, one with his son Wilhelm and the other with Alexander von Humboldt when the optimal strategy for the geomagnetic measurements was discussed. Wilhelm's mild and pleasant nature helped to avoid an open confrontation but his father clearly resented his independence and stubbornness. Their relation was never stable until Wilhelm emigrated and settled in Louisiana to start a new life as a farmer; in 1838, Gauss sr. was pleased and proud to be able to report to Olbers the birth of his first grandchild, who was a citizen of the new world.[2]

While Gauss was reluctant to jeopardize his peace of mind in another conflict with one of his sons, he vigorously fought against Humboldt's designs for magnetic experiments, primarily in the correspondence with Schumacher.[3] The point is not whether Gauss was right or wrong—he was essentially right—but rather that some details of the design of experiments were of so much importance to Gauss that he openly and with much emotional involvement attacked a man who had been (and continued to be) an admired friend and prestigious colleague.

Though he was not exclusively involved in experiments, they now were Gauss's strongest scientific interests. Interested as he was in other areas of human life and activities, he appeared to withdraw himself from certain sectors and was careful not to be distracted or to be drawn into matters which would demand an inordinate amount of his time. Gauss was never politically active and even refrained from making explicit political comments. It is difficult to tell what his political and, more generally, his philosophical opinions and convictions were, but one can get at least some vague notion from his reactions and occasional remarks in the correspondence. An obvious

Gauss, approximately 55 years old (sketch by J. B. Listing; nachlaß Gauß, Posth. 26). Courtesy of Universitätsbibliothek Göttingen.

first impression is that of extreme political opportunism. Gauss never dissociated himself from his youthful experiences when the generosity of his prince allowed him to become a scientist and helped him to leave behind the narrow and hopeless world into which he had been born. Gauss, of course, was able to see the limitations and shortcomings of the feudal system—they were too blatant to be overlooked; equally obvious, he was himself the beneficiary of reforms which were direct or indirect consequences of the French Revolution. His view of the time in which he lived was a curious mixture of conflicting sympathies. There was Gauss's gratitude to the old order, coupled with some sort of old-fashioned German patriotism. (His rejection of many of the new political developments was motivated by the fact that they came from France and in conjunction with the humiliating Napoleonic wars.) Another factor in Gauss's political attitude, with conflicting consequences, was the healthy and firm assessment which he had of himself. His self-assurance increased with age, but it was surprisingly strong even at the outset of his career. He was the acknowledged Prince of Mathematicians, an appellation which he easily accepted.

Gauss's basically aristocratic attitude expresses itself openly and freely in this regard. Examples are his evaluations of the work of others, or his demands for instruments and the funding of costly experiments. Gauss made his requests without false modesty, was forceful whenever necessary, and occasionally left his government little choice but to follow his advice.[4]*

All these attitudes are quite incongruous and do not lead to a picture of a unified and consistent character. It appears that different strains of behavior prevailed in different situations; what is manifest from his "public" reactions might have expressed itself quite differently in personal social life or in private opinions.

We have already seen what Gauss's official attitude towards his government was, as a citizen and as a scientist. It was determined by his initial gratitude and a strong desire to be a good and useful citizen. Towards his colleagues, Gauss was largely indifferent—he was too certain of his superior genius to discuss or to expect much. This same conviction seems to have influenced his attitude in the conflicts with his family, and though there were also other psychological mechanisms at work, these factors are of no interest to us. Gauss's politics are usually described as conservative, for the reasons sketched above, but also because of his anti-French sentiments, which he never tried to hide. There was also his inactivity during the protest of the

---

* This reaction, more than anything else, helps to explain Gauss's decision to reject the call to Berlin. Prudently (from his point of view) Gauss was not directly involved in the negotiations, but he must have felt that the bartering and bickering about his salary and other details of his position were very humiliating. Ultimately, the generosity and swiftness of the aristocratic Hanoverian government won him over. In this sense, Kummer was right when he says that Gauss did not come to Berlin because of a few talers' difference.

"Göttingen Seven" in 1837/38, a reaction that will be discussed below, and a common misunderstanding by later generations of what German nationalism and conservatism meant during this period.

The picture of Gauss's incongruous behavior which we have tried to project should not come as a surprise even to us, educated as we are to believe in the virtues of a strong, well-rounded, unified character. There are, of course, major themes under which Gauss's life can be subsumed. Dealing with them would immediately lead to complicated and fruitless discussions of the historical situation. Such discussions are superfluous because it is very doubtful that the era in which Gauss lived allowed the formation of a unified personality for one in his station and with his upbringing. Gauss's outlook, in today's terms, was fragmented or even anarchic but there was a strong (and for the sake of self-preservation, indispensable) streak of egoism in his character. In addition, there was Gauss's desire to be left alone, to avoid problems and conflicts. We have already seen this attitude during his conflicts with his family, and we will encounter it again in the course of the political confrontation of 1837/1838.[5]

Among Gauss's favorite contemporary writers were Walter Scott and, of the Germans, Jean Paul, a popular novelist. In the light of what was said above, Gauss's interest in contemporary literature may be surprising, but not if one bears in mind that both Scott and Paul studiously avoid contemporary problems. This is particularly true of Jean Paul's later novels; his work can be interpreted as developing from revolutionary activism to quietistic idyllism, eased and mitigated by a sharp ironic humor of dubious motivation.[6] Jean Paul's approach obviously satisfied the needs of his audience, whom he provided with a detached and (on the surface) complacent interpretation of the turbulent changes which swept over the country and radically changed the political and social life. Gauss was quite uninterested in the products of the so-called classical and romantic schools of German literature—both the emotional enthusiasm of the latter and the upper middle class elitism of the former were quite foreign to his world and experiences.*

---

* Jean Paul's background is similar to Gauss's though there are important distinctions in their development. Coming from a humble background, Jean Paul was poor, nearly destitute, during the first 30 years of his life. He earned his living as a schoolmaster, and his first publications were not successful. They were politically progressive and influenced by Rousseau and the French Revolution. Later, Jean Paul gave up activism; through his idyllic and scurrilous novels, he became the darling of the literary salons and educated public.

CHAPTER 12

The Göttingen Seven

Politically, times had been quiet following the defeat of Napoleon and the establishment of a European security system by the Congress of Vienna in 1814/15. One of the first major upsets was the bourgeois French revolution of 1830 during which Charles X was deposed and replaced by Louis Philippe. The unrest in France had repercussions in several other European states and led to the establishment of an independent kingdom in Belgium which had previously been part of the Netherlands. As in other German towns, there was some restlessness in Göttingen but it did not affect Gauss much. He, of course, did not sympathize with the unruly students and their activism which, he thought, could never succeed.[1]

Gauss's assessment turned out to be wrong. The protests in Göttingen and other parts of the kingdom were taken seriously by the governments in Hanover and London. Without much pressure, a new constitution was granted which was much more liberal and democratic than its predecessor. This brought Hanover in line with the majority of the German states after being, despite its affiliation with England, the reactionary taillight in the *Deutscher Bund*, the confederation of German states which had replaced the Holy Roman Empire.

This new constitution triggered the crisis of 1837/38 and led to important changes in Gauss's life. In 1837, King William IV of England died without legitimate offspring. In England, Queen Victoria succeeded, but the succession in Hanover, regulated by Salic law, did not allow for a female ruler. This ended the union between England and Hanover, and one of the Queen's uncles, the Duke of Cumberland, became the new King of Hanover.*

The crisis came quickly, though not before there was time to celebrate the centenary of the university in the presence of the new King. There were

---

* His shadow over the English throne was the main reason for the general sigh of relief with which the birth of Princess Victoria had been greeted in England. Duke Ernest (August) was the least appealing of the generally unappealing sons of George III. The new King was thoroughly reactionary, and his accession in England might have led to a crisis in which the institution of the monarchy itself might have been questioned. Compare, for example, Lytton Strachey's biography of Queen Victoria [London 1921].

speeches, services, and the distribution of honors; reports of the festivities tend to emphasize the feeling of impending doom which accompanied the celebration. Shortly after the jubilee, the King declared the constitution void and annulled the oath which had been sworn to it by the civil servants in his country, among them the professors at the university in Göttingen. The reasons for the King's action are not quite clear, but there was at least one clause in the constitution which he found intolerable. It automatically barred from the throne any prince who had a major physical defect. The King's only son was blind.[2]

Ernest August's high-handed action immediately led to a confrontation with his people, who had become more and more assertive; the university quickly turned into one of the centers of protest. Seven professors, including Ewald, an orientalist and the husband of Gauss's second child Minna, and Wilhelm Weber, signed a formal protestation, declaring that the King's action could not release them from the oath that they had sworn to the constitution of 1831. All seven lost their positions, and the activists among them, not Ewald or Weber, were forced to leave the country overnight. A wave of protest swept through Germany, and all seven received offers of new positions at other universities. Weber returned to his native Saxony, to Leipzig, and Ewald eventually accepted a call to Tübingen, in southern Germany, after doing research in London for a couple of years. Even Prussia was not idle and gained the eminent Grimm brothers for Berlin and the law professor Albrecht for Königsberg.[3] Ernest August assessed the situation calmly, remarking that he could hire university professors as easily as ballet dancers.

Gauss's apparent inactivity during the conflict has been interpreted in different ways, as an expression of his conservative leanings or as a betrayal of his friends and colleagues. Originally, the protesters had hoped that he would join them; his voice would have given their protest a wider echo and much more effect. But this could never realistically have been expected, regardless of what Gauss's political convictions were. He was not prepared to expose himself and to lend his prestige to a cause so far removed from his primary concerns. The loss of Weber was clearly irreparable, but in Gauss's eyes even this could never justify using his authority in an extraneous and inappropriate way.

The correspondence with Schumacher gives a good idea of Gauss's attitude during the conflict. Schumacher was not really a congenial correspondent in this matter. He was much more conservative, in a modern sense, than Gauss was and made several explicit statements about the folly of the protesters; Gauss is much more circumspect in what he says. Their discussion is not about the issue itself, but rather how the accompanying disturbances affected Gauss. Schumacher was concerned that Gauss was about to leave Göttingen and accept a position in Paris; we quote from Gauss's answer: *I hasten, my dearest friend, to answer your letter of Dec. 8 which I received only today. The newspaper article about me (. . .) is but one of the many lies which fill today's public papers: I have never told anybody what I do or do not intend to do.*

I wish and hope that the university as a body will not enter the political scene. You know that two persons who are very close to me were drawn into it by assenting to sign the well known protestation. The investigation, now pending before the university court, only concerns, if my information is correct, its unauthorized propagation, and the two men were not in the remotest way involved in that. I therefore cannot believe that their signature will have unpleasant consequences, and as long as these two strong magnets are undamaged in their places, Göttingen will be much more attractive for me than Paris. Whether I would ever, in circumstances which would make me leave Göttingen, prefer Paris to other places, is a question which need not be discussed now. . . .[4]

Gauss's hope that Ewald and Weber could stay was wishful thinking; the next letter to Schumacher, written a week later, accepts the inevitable and contains the data of an observation of the star 57 δ Arietis. No mention was made of any intention to leave Göttingen. Gauss, however, did try to intercede indirectly for his two friends, but without success. The King's conditions for their reinstatement were too humiliating to be acceptable, and it must be mentioned that Gauss never prodded Ewald and Weber to recant explicitly or to act against their convictions. One should not forget that Gauss was sixty years old when the crisis occurred and that a full generation separated him from his younger colleagues.[5]

Gauss's actions during the constitutional crisis do not automatically make him a political conservative, and there were other situations in which he appears in a different light. He was encouraging and interested when the University of Marburg sent Gerling as its representative to the local parliament in Cassel in 1833[6]; later, when he was quite old, he did not show any sympathies for the revolution of 1848/49, but he tried to protect Eisenstein and Jacobi, who had both been active in the liberal cause.[7] Nor did Gauss share Schumacher's antisemitism, though he did not rebut Schumacher's invectives, even when they concerned Jacobi, whom Gauss regarded highly. Altogether, Gauss appears to have been remarkably free of personal prejudices of this kind—even his hatred of Napoleon and of French dominance in Europe did not interfere with his appreciation of his French colleagues, or develop into a rejection of everything that came from beyond the Rhine.

The Method of Least Squares

As we have just seen, the period of unexpected productivity and fruitful cooperation with Wilhelm Weber came to a close with the end of the year 1837. Gauss and Weber continued their joint work as the editors of their geomagnetism journal but their separation did not permit any genuine exchange of ideas. In 1849, Weber returned to Göttingen, after yet another political upheaval, but by then Gauss was very old and no longer active in research. The last 18 years of Gauss's life were free of the painful commotions which had accompanied and marred his earlier years, but they were also no longer productive in a strict sense, though Gauss was far from inactive. Below, we shall give an account of Gauss's scientific work during this period, but it seems appropriate to pause for a moment and discuss in what light Gauss viewed his own research and the philosophical notions which accompanied it. They assumed their final shape during this period of Gauss's life and were now much clearer than in his earlier approaches.

One of Gauss's most efficient tools in his research was the method of least squares. When he first developed it, shortly before the turn of the century, he did not consider it very important; Gauss remarked later that he was certain that his predecessor at the astronomical observatory in Göttingen, the elder of the Tobias Mayers, had known the method. After looking through Mayer's papers Gauss saw that this was not the case, but he still could not take credit for the invention of the method. Formal priority belongs to Legendre, who first published it in 1806, i.e., clearly before Gauss made the method public, though he had clearly and frequently used it much earlier.[1]

Gauss motivated and derived the method of least squares in several substantially different ways. Here, we summarize his most "mature"[2] approach, as developed in the two papers "Theoria combinationis observationum erroribus minimis obnoxiae" I and II (1821, 1823). Part I is devoted to a systematic and theoretical investigation of the theory of errors, presented as a part of probability theory. Of the two, essentially different, types of error, systematic and accidental errors (the latter are the *Zufallsfehler*), only the accidental errors are relevant; for certain domains, a certain probability can be assigned to them. Formally, Gauss defines the function $\varphi(x)$ as the relative error in the observation x. Then $\varphi(x)dx$ expresses the probability of an error between x and $x + dx$. φ is normalized by the condition

$$\int_{-\infty}^{\infty} \varphi(x)\,dx = 1.$$

The decisive requirement is that the integral

$$\int x^2 \varphi\, dx$$

attain a minimum. This condition expresses the idea that the square of the error is its most suitable weight. This is where Gauss's approach is different from that of Laplace, who earlier had tried to use the absolute value of an error for its weight. This is why Gauss's method is called the method of least squares: computationally, it is clearly superior to Laplace's original method.

After developing the theoretical basis of his theory, a suitable function φ had to be found. In general, the distribution of the errors will not be known in advance, and one has to choose from arbitrary functions ψ with the single requirement that $(*)$ be satisfied. At this point, Gauss introduces, after some heuristic preparations, the exponential e^{-x^2} ("normal distribution") as a particularly natural way in which the errors of observation occur. Gauss concludes Part I of the paper with some complicated considerations which are motivated by astronomical questions and are of no interest here.

The second part of "Theoria combinationis . . ." contains applications of the method of least squares, mostly to problems from astronomy. Gauss also develops a complicated, but not difficult, elimination procedure to determine the best observations. Another problem concerns the inclusion of new experimental data at an advanced stage of the evaluation. Gauss shows how to make use of the new data without having to discard any earlier calculations. The paper ends with discussions of the relation between the observed and the calculated errors and of the accuracy with which the average error can be determined.

A supplement is devoted to a problem from geodesy. It deals with a situation in which it is not known how the observed parameters depend on certain other, equally unknown, elements; only their mutual dependence is given explicitly. Gauss solves this problem by deriving once again the method of least squares; the approach is basically the same as his earlier derivation. The supplement also contains two worked out examples from geodesy in which Gauss uses his own data and data from the Dutch triangulation; he does not omit to comment how important the use of real data is. The direct measurement of the three angles of a triangle will usually not add up to 180°.

Gauss never mentioned, in any of his papers, the possibility of statistical distributions other than the normal one. The very satisfactory results which he obtained by using the exponential distribution did not prompt him to look for other approaches.

Gauss gave altogether three different derivations of the method of least squares, the first of them in *Th. mot*. Legendre, in his brochure of 1806 on the orbits of comets, seems to have developed it from a strictly computational motivation, and this appears to have also been Gauss's initial approach. Soon after Legendre had published the method, Laplace succeeded in connecting the least squares with probabilistic considerations, but we do not know whether Gauss used Laplace's work or whether he developed the foundations independently. There are some reasons which make the latter assumption more likely; in any event, Gauss's work went beyond Laplace's, and for Gauss the method gained in value by its probabilistic justification.

Least squares were Gauss's indispensable theoretical tool in experimental research; increasingly, he came to see it as the most important witness to the connection between mathematics and nature. Its efficiency was the most blatant demonstration of the fact that natural phenomena could efficiently be investigated by mathematical methods. Gauss would have expressed this fact by a much stronger statement—for him, mathematics governed the workings of nature, and the mathematical penetration of the natural sciences showed to what degree they had been understood.

We have seen that there were other mathematical techniques which Gauss found helpful in his desire to understand natural phenomena and processes. They were potential theory, including Coulomb's law, extremal principles, and the calculus of variations. Gauss was even aware of an application of the theory of numbers: he knew that crystalline structures could be described with the help of ternary forms.[3] These examples fortified Gauss's conviction of the mathematical character of nature; this belief had its roots in the 18th century, but Gauss did much to make it more credible.

There are few explicit remarks by Gauss about his understanding of nature and the role of mathematics in the physical sciences. We quote here the last paragraph of his paper about an extremal principle in mechanics:

It is very remarkable that the free movements, if they cannot coexist with the necessary conditions, are modified by Nature in exactly the same way in which the calculating mathematician, according to the method of least squares, reconciles observations which are connected to each other by necessary dependencies. This analogy could be pursued further, but I do not intend to do so at the moment.[4]

Volume XII of Gauss's works contains a lengthy article with the title "Astronomische Antrittsvorlesung" ("Astronomical Inaugural Lecture"). It cannot be dated exactly, but it is quite early. In it, Gauss discusses a variety of topics which one might mention in such a lecture, most importantly the different motivations for the study of astronomy. Astronomy is a useful science, but Gauss sees this only as a secondary motive. It is primarily the disinterested quest for truth which makes astronomy such a gratifying object of research. The highest reward of the astronomer is the satisfaction of being able to contemplate the wonderful ways in which the world has been organized and to experience the reassurance which comes from the recognition of the harmony of the creation.*

One can find out more about Gauss's opinions and beliefs from remarks in the correspondence. Occasionally, he ventured philosophical statements, expressing a general disdain for the vagueness of most philosophers. There are derogatory remarks about Plato, Wolff, and the contemporaries Schelling and Hegel.[6] Gauss used Hegel's absurd astronomical speculations in his dissertation as an example of the stupidity of his kind. Kant and his work were appreciated, though naturally not his geometrical ideas (Kant proved the "necessity" of Euclidean space) and, more importantly, his a priori classification of mathematics as a synthetic science. We also know that Gauss was quite fond of the work of J. F. Fries (1773–1843), one of the few philosophers of this age with a serious interest in the experimental sciences, particularly astronomy. Philosophically, Fries stood in the tradition of Kant, but tried to integrate the contemporary elements of positivism into his philosophical system. Gauss was particularly fond of a history of philosophy which Fries wrote.[†7] In general, Gauss was not interested in philosophical arguments, and his attitude reflects more the modern scientist than the philosophically inclined mathematician or scientific scholar of the 18th century.

---

* In this lecture, Gauss makes a "political" statement which may come as a surprise. He argues against a narrow utilitarian point of view; we quote:

To judge in this way demonstrates not only how poor we are, but also how small, narrow, and indolent our minds are; it shows a disposition always to calculate the payoff before the work, a cold heart and a lack of feeling for everything that is great and honors man. One can unfortunately not deny that such a mode of thinking is not uncommon in our age, and I am convinced that this is closely connected with the catastrophes which have befallen many countries in recent times; do not mistake me, I do not talk of the general lack of concern for science, but of the source from which all this has come, of the tendency to everywhere look out for one's advantage and to relate everything to one's.physical well-being, of the indifference towards great ideas, of the aversion to any effort which derives from pure enthusiasm: I believe that such attitudes, if they prevail, can be decisive in catastrophes of the kind we have experienced.[5]

This last allusion seems to refer to the Napoleonic wars and to the defeats of the German states and Prussia. The latest date for the lecture is 1815 (the year of Napoleon's final defeat) but it could have been as early as 1808. If given so early it would have been a bold and explicit political statement.

† Fries, never particularly popular, was forgotten soon after his death. In the 1920s, the philosopher Leonard Nelson (1882–1929) started a Fries renaissance in his attempt to find some new direction between Husserl's phenomenalism and the then vigorous neo-Kantian school. Nelson, a grandson of Dirichlet, taught at Göttingen and was quite popular among the mathematicians of the Hilbert school.

CHAPTER 13

Numerical Work. Dioptrics

The previous interchapter studied Gauss's view of the role of mathematics in the natural sciences. Now, we turn to an "intramathematical" topic, his fondness for and proficiency in numerical calculations.

It would not be correct to consider Gauss's extensive numerical work as a waste of time, unconnected with his theoretical work, an uninteresting distraction forced on Gauss by the economical and social circumstances. Gauss's numerical work is part of his "theoretical" considerations, often providing first clues for his discoveries and conjectures. Gauss's calculations were very extensive, and he may have performed more of them than any other eminent mathematician with whose work we are familiar. An enormous number were necessary for the reduction of his experiments; indeed, Gauss's reputation and efficiency as a scientist cannot be separated from his seemingly unlimited capacity to *reduce* the raw data of his observations and to refine them with the help of the least squares. The theoretical side of the numerical work has often been underestimated, perhaps as a consequence of a recent emphasis on "rigor" and nonnumerical arguments. That this could have happened is yet another consequence of Gauss's own work and that of the subsequent generations of mathematicians who worked under his influence. Only at the beginning of this century were quantitative considerations of the kind Gauss used put on a firm and generally accepted basis, when numerical analysis was systematically developed. Now, after the advent of electronic computing devices, the field is again of a substantial intramathematical importance. For Gauss, numerical matters were an unquestioned and genuine part of mathematics. What distinguishes his work from that of his contemporaries and predecessors is the critical evaluation of the results and a strict distinction between heuristic ("inductive") and rigorous derivations. A case in point is Gauss's use of convergence criteria for infinite series. Convergence is actually a topic in the "theoretical" paper about the hypergeometric function, but is not discussed elsewhere, not even in the context of Gauss's many summations of infinite series and calculations of estimates. In these cases, convergence must have appeared to Gauss to be either unimportant or obvious.

The number of calculations which Gauss had to perform for his mathematical and scientific work can hardly be imagined today. One has to bear in mind that only logarithmic and a very few similar tables aided his extensive calculations. Slide rules and other mechanical devices did not exist.[*][1]

Essential as Gauss's numerical versatility was for his research in applied mathematics, pure mathematics seems to have profited more from it. Most notably, Gauss's number-theoretical work is deeply and directly rooted in numerical considerations. These considerations go back to his youthful, seemingly aimless play with the natural numbers. Examples of work that is particularly close to its numerical roots are the first proof of the law of quadratic reciprocity, and, more typically, the connection between the arithmetico-geometric mean and elliptic integrals. Examples like these do no more than illustrate our point; they are not accidental. At each step, Gauss's research was close to the "numerical reality"; it was guided by his deep knowledge of the natural numbers.

Gauss was particularly fond of tables which could be used to simplify lengthy calculations. *Disqu. Arithm.* contains several important tables, as do many of Gauss's other works. In his book reviews, Gauss liked to emphasize the value of such tables and to discuss their efficient organization and design.[2]

In an essay in Vol. X of the collected works, P. Maennchen analyzes the interplay between Gauss's theoretical and numerical work, emphasizing the theory of numbers. It is in this area that Gauss made his best-known conjecture. This was the prime number theorem, a result which he found very early in his career as an empirical distribution.[3] As a consequence of Gauss's incessant and assiduous calculations, many numbers possessed for him individual characters, almost as if they were living things. He was able to exploit nonobvious arithmetical properties to simplify and shorten his calculations, calling the resulting techniques, tricks nearly, his *artificia*.

Despite the *artificia* and his great experience and proficiency, Gauss's calculations are by no means error-free. Errors, most of them minor, are quite common in his reductions of astronomical and geodetic measurements; strangely enough, Gauss often carried out calculations to more decimal places than was reasonable for his experiments. Another contributing factor was that Gauss never controlled his results or checked his calculations— whatever his motive was, he again appears to be rather the observing mathematician than the calculating observer.

If one looks through Gauss's notebooks and the fragmentary material which was posthumously published in the later volumes of the collected works, one sees how much he enjoyed physical manipulation of numbers and immersion in vast calculations. Sometimes only this, rather than actual interest in the result, appears to have been Gauss's principal motivation.

---

* Gauss was interested in calculating machines but did not take part in the contemporary development of the first prototypes. He seems to have met Babbage in Berlin when he visited Humboldt, but we don't know of any comments on Babbage's machines.

Gauss discussed in the correspondence the feats of several of the phenomenal "master calculators" who showed their talents in public exhibitions.[4] He was not very impressed, because they appeared to work from memory, without the use of notable *artificia*. When one of them, Z. Dase, asked Gauss how he should use his talent, Gauss advised him to extend the existing tables of factors. Dase followed this suggestion and later published useful tables for the numbers from 7,000,000 to 9,000,000.

If the calculations were the most fundamental and "disinterested" part of the work of Gauss, providing equally a basis for his theoretical and for his applied investigations, it was his practical work in optics that was the most specialized area of empirical research in which Gauss ever worked. His *dioptrical* investigations, as he called them, were concerned with the shape, the arrangement, and the defects of optical lenses. Gauss's involvement in this area was quite natural; in fact, every observing astronomer of the time had to contend with problems originating in the unsatisfactory quality of the available glass and the technical limitations in cutting and grinding it, and insufficient knowledge of the theory of telescopes. All these factors changed during Gauss's lifetime, and Gauss himself contributed much to this development. The biggest deficiency which affected Gauss's work was never taken into account by him at all: Gauss approached all optical questions on the basis of a naive corpuscular theory of light and did not show any interest in Fraunhofer's theory of the propagation of light.[5]

Gauss became active in dioptrics in 1807 when J. G. Repsold, a renowned instrument maker from Hamburg, asked him about the lenses for an achromatic double objective.[6] Gauss immediately involved himself in the question though problems of this kind were new to him. The exchange with the Repsolds, father and son, continued for many years, but Gauss soon started to go his own way and to pursue some independent research. His most notable work concerned lenses with a minimum of chromatic aberration. He discussed his problems with other astronomers, among them Bessel and Olbers, and with instrument makers, among them the famous Utzschneider, Fraunhofer, and Steinheil who all worked in Munich but in different shops. To follow Gauss's work, one has to study his correspondence and certain textbooks of astronomy which originated from his lectures.[7] There are also a few short papers which Gauss himself published. His work in dioptrics had a considerable influence on the development of the optical industry in Germany* but is obsolete today because of Gauss's assumption of the corpuscular nature of the propagated light. Gauss's most important optical

---

* In the course of the 19th century, Germany emerged as the world's leading supplier of fine optical instruments, replacing England. During the first half of the century, the shops of Reichenbach, Fraunhofer, and Steinheil were established; the most famous of them, Carl Zeiss in Jena, was founded several decades later. Ernst Abbé, the scientific director of the *Zeiss-Werke*, happened to be a student of Riemann.

paper is *Dioptrische Untersuchungen*, published in 1840. It is a detailed study of the path of a ray of light through a system of lenses of nonnegligible thickness. Gauss's main result is the reduction of such a complex system to a single, infinitesimally thin lens. Even here, we see how Gauss reduces the problem to a mathematical one and develops a mathematical theory rather than a sequence of technical instructions or a textbook of physics.

From the correspondence, we mention a curious discussion of the effects of nearsightedness for the user of a telescope. Gauss, nearsighted himself, comes to the (incorrect) conclusion that nearsightedness would result in an additional magnification for the observer.[8]

Essentially, Gauss's investigations continued the work of Euler and Laplace. They do not dramatically break new ground but clarify what was known, and they keep abreast of the technological development of his time. They are, as we saw, not isolated with Gauss's work and show not only his scientific interests, but also his willingness to cooperate with nonscientists and to share his theoretical and practical insights with them. Mathematically, *Dioptrische Untersuchungen* is so elementary that Gauss was reluctant even to publish it.

It would be wrong to assume that Gauss was only interested in the theoretical side of the matter. He was always involved in changes of the setup of his instruments in the observatory and discussed with his friends much time-consuming manual work, such as what the best way was to draw threads through the eyepiece of a telescope. Beyond this, Gauss showed genuine and original interest in technological progress; one of the most enjoyable trips of his life took place in 1815 when he visited the different workshops in Munich and vicinity in order to discuss the equipment for his new observatory in Göttingen.

CHAPTER 14

The Years 1838–1855

Our picture of Gauss, as it has come down to us from the years between 1838 and 1855, is even paler than that from the preceding era. There are now several eyewitness reports from students and vistiors, but even they do not contribute much.[1] Most of the letters from this period were to Schumacher, now by far his most frequent correspondent, but they stay on the surface, projecting the picture of a perpetually busy man with many diverse interests, most of them apparently not very inspired. Gauss was still quite active in his astronomical and magnetic observatories, but there were many other occupations. There were elementary mathematical problems, among them combinatorial ones posed by Schumacher,[2] experimental and theoretical physics, and once more foreign languages.

There are quite a few surprises even in the Schumacher correspondence which show us Gauss in an unexpected new light. In a letter of December 1824 (#228a of Dec. 23), he pleads with Schumacher not to dismiss his young assistant Klausen even though he was clumsy and had dropped a precious barometer. Gauss asks Schumacher to be tolerant and writes

. . . I hope you will accept him again but I am concerned about his future. The incident with the barometer may not have been the first of its kind and lets me believe that he is not skillful enough to be a practical astronomer. This and teaching, however, are currently the only ways which allow a mathematician without means of his own to earn his living. Only if he were to produce something exceptionally good could he hope to find a position at an academy, the way they are organized today, and even then you could bet 99 to 1 that he would not get it. I do not know whether he could ever be a professor; you are the better judge. If you think he would also not be suitable for this then he would perhaps do best to go into a different profession, he could join the army or do something else where he could use his spare time for mathematics. If one has to work for one's living it does not matter whether one teaches beginners or works as a cobbler. The only thing that counts is what would leave a maximum of free time . . .[3]

This passage is typical for Gauss's attitude during the last 30 or 40 years of his life. He strove to be mild and understanding and though he could not always maintain it, this seemed to be a genuine expression of the way he saw things during this period.

In the early 1840s, Gauss took up Russian after a short flirt with Sanskrit, which he did not find very congenial. He maintained his interest in English literature; a triumph, immediately reported to Schumacher, was the identification and correction of the sentence "The moon rises broadly in the northwest" in one of the novels of the much loved Walter Scott.[4] Gauss's concern for details should not come as a surprise. He had an indefatigable interest in factual information whether it was important or not. We recognize a related trait in his mathematical investigations: they were always inductive, proceeding from particular facts to general statements and avoiding unnecessary abstraction. It was this which led him to disregard any algebraic notions in *Disqu. Arithm.*, though he developed them in the course of the exposition and only stopped short of defining them explicitly. So far, we have stressed in this biography the mathematical character of Gauss's work in the natural sciences, but one should not overlook the "scientific", often experimental character of his mathematical work. Gauss's thinking was inductive to an extraordinary degree, hence his hunger for facts, his love of details, be it in mathematics, in the natural sciences, or in any other sector of his intellectual life. This is the light in which we have to see his untiring discussions of astronomical or geodetic measurements and his experiments on behalf of the Hanoverian commission on weights and measures, as well as his endless computations, his interest in tables of prime factors, or his computational determination of one of the periods of the lemniscate.

In 1842, though Gauss had not spent a single night out of his house for over ten years, he entered into negotiations with Vienna, where a position at the university had been offered; if his Berlin negotiations, 20 years earlier, had not been very purposeful, the short negotiations with Vienna were even less so and broke off quickly.[5]

The years following Weber's departure were deeply unhappy. Gauss was spared the extreme humiliation which often accompanies old age, but it would be deceptive to try to project the picture of a serene evening in a productive and fruitful life. The loss of Weber in early 1838, who in so many respects had been like a son to Gauss, followed the emigration of Wilhelm; shortly before Weber, Minna Ewald and her husband had left Göttingen. In 1839, Gauss's old mother, now blind, died; her granddaughter Minna followed in 1840, only in her 33rd year. This last loss was very painful—she and Joseph had been the surviving children, the "pledges", of the first marriage. Minna, her father's favorite among the children, was said to have resembled her mother closely. Of the friends, the revered Olbers died in 1840,

Bessel in 1846. The correspondence with the latter had again been frequent, but without the old intimacy. Occasionally, Gauss mentioned his physical ailments: there was a temporary, unexplained deafness in 1838, increasing loss of memory and vision, and the loss of teeth. Still, despite all this, the mechanical duties of the observing astronomer and his diverse other interests gave Gauss the distraction he needed.

It is not possible to discern a radical difference in Gauss's behavior between the early and the late 1830s. Outwardly, he remained unaffected by the painful experiences he had to undergo; what we observe is a steady decline, presumably painful for Gauss himself to watch. Not much else was indeed to be expected—Gauss had learnt not to look for happiness. Friendship and scientific collaboration were replaced by personal honors, medals, and other distinctions. Jacobi, Dirichlet, and Eisenstein, perhaps the most gifted German mathematicians of the two generations after Gauss, paid their respects, but they did not establish lasting or personal contacts.*[6] In 1838, Gauss received the Copley medal from the Royal Society in London; later, he was made a member of the order *pour le mérite* in Prussia.

Of the immediate family, only his daughter Therese stayed in Göttingen. She was unmarried; in 1838 she took over the management of her father's household. Though Gauss was very fond of her it does not seem that father and daughter had much in common, except a strong bond of mutual appreciation, thankfulness from the father's side and admiration from the daughter's.[†]

Reports from several of his students who attended his lectures when he was an old man show that Gauss now enjoyed teaching and lecturing much more than in his early years.[7] He clearly was a good and competent teacher. One of the reasons for his change of attitude may have been that students were now much better prepared and more interested. Among Gauss's last students were G. Cantor, who is known as the author of a voluminous history of mathematics, and R. Dedekind, famous for his work in number theory and algebra. We quote from an account which Dedekind gave in 1901 (translated in [Dunnington]):[8]

. . . usually he sat in a comfortable attitude, looking down, slightly stooped, with hands folded above his lap. He spoke quite freely, very clearly, simply and plainly; but when he wanted to emphasize a new viewpoint, in which he used an especially characteristic word, then he suddenly lifted his head, turned to one of those sitting next to him, and gazed at him with his beautiful, penetrating

---

* Gauss followed their work and careers with much interest. For Eisenstein, he wrote a very complimentary preface to a collection of his number-theoretical papers.

† There is an ironic postscript to Therese's story. After her father's death she first was inconsolate. In 1856, only a year after her father died, she married a theatrical producer and actor with whom she had entertained a romantic correspondence since the early fifties. This step was frowned upon by other members of the family, certainly by Joseph, presumably because it was considered a mésalliance which desecrated the family name and the memory of the great man.

*blue eyes during the emphatic speech. . . . If he proceeded from an explanation
of principles to the development of mathematical formulas, then he got up, and
in stately, very upright posture he wrote on a blackboard beside him in his
peculiarly beautiful handwriting; he always succeeded through economy and
deliberate arrangement in making do with the rather small space. For numerical examples, on whose careful completion he placed special value, he brought
along the requisite data on little slips of paper.*

If one goes through the list of courses taught by Gauss one sees that most
of them were in astronomy and only very few in number theory and other
areas of pure mathematics. He lectured quite often on the method of least
squares and its applications in science. In fact the only time Gauss ever taught
number theory was in 1807/08, his first year as a professor in Göttingen.[9]

Gauss's involvement in concrete experimental questions dates, as we saw,
back to the beginning of the century. It led Gauss to cooperate with other
astronomers; most of the friendships of his adult life have their roots in joint
astronomical work. Gauss never forgot the other side that went with it—that
this involvement precluded a deeper involvement in pure mathematics, his
always-remembered first love.

In 1838, the year Weber had to leave Göttingen, Gauss completed the 61st
year of his life. He had lost the only colleague in Göttingen with whom he
had ever been able to establish a genuine and mutually fruitful working relationship. There was no hope of adequately continuing their cooperation, and
yet Gauss was able to go on with his observations without interruption, to
contribute to the collection of geomagnetic data which was being compiled,
to correspond with Schumacher, Gerling, Olbers, and Bessel about astronomical and other questions. We see the beneficial aspect that the practical
work had: Gauss went on to make useful scientific contributions at an age
when his original mathematical productivity would perhaps no longer have
held up, and in the face of the various personal disasters which he had to face
after 1809. Not only did he not lose his ability to work and to make substantial scientific contributions, but his work actually helped him retain his equanimity and avoid frustrating and fruitless confrontations. Above, we saw
the other side to this—Gauss's unresponsive and inflexible attitude in the
conflicts with his sons, and the end of concentrated mathematical research
soon after 1810, but these sacrifices may have been necessary if a continuation
of his work was to be possible at all. So the mechanical, the automatic
component of the experimental work contained an enormously positive, sal-
vaging factor and should not merely be seen as an unwelcome distraction
from more creative and "valuable" activities.

What was said above should not indicate that Gauss completely gave
up his interest in higher mathematics. Even in the last fifteen years of his
life, twenty or thirty years after his last original research, there are still
many interested and interesting remarks and observations in the correspondence. Gauss found the time to involve himself quite seriously in two
topics in which he had had a strong interest long before. He was among

the first mathematicians in western Europe to understand and appreciate Lobachevski's research on non-Euclidean geometry. The way Lobachevski presented his results is quite difficult, even obscure; his earlier papers, including those which appeared in German, were without immediate resonance.[10] In 1846 (Nov. 28), Gauss had occasion to make the following remarks in a letter to Schumacher:

Recently I had reason to look again through Lobachevski's booklet (Geometrische Untersuchung zur Theorie der Parallellinie. G. Funcke, Berlin 1844. 4 Sheets). It contains the elements of a geometry which would hold, and which could rigorously hold, if the Euclidean would not be the true geometry. A certain Schweikart called such a geometry astral geometry, Lobachevski calls it imaginary geometry. You know that I have had the same conviction for 54 years (since 1792), with a certain later extension which I do not want to go into here. There was nothing materially new for me in Lobachevski's paper, but he explains his theory in a way which is different from mine, and does this in a masterful way, in a truly geometric spirit. I thought I should draw your attention to the book, it will provide you with exquisite pleasure.[11]

This passage, interesting for several reasons, contains the apparently erroneous statement that the possibility of the existence of non-Euclidean geometries had been clear to Gauss when he was 15 years old.

It was in connection with Lobachevski's work that Gauss started his Russian studies. He did not confine himself to mathematical texts, and complained in a letter to Schumacher that the selection of Russian literature in the local library was quite meagre. His favorite poet seems to have been Pushkin.[12]

In 1849, the University of Göttingen celebrated the 50th anniversary of Gauss's doctorate, his *golden jubilee**. The main event was a ceremony in which Gauss submitted an improved version of his doctoral dissertation, again treating the fundamental theorem of algebra. The changes from the earlier version do not go very deep; the most notable difference is the explicit use of the complex domain which had not been mentioned at all in the original dissertation of 50 years previously, though Gauss had "implicitly" made use of it. The paper shows how firm was Gauss's grasp of the subject matter— he was still an active and competent mathematician. Mathematics was still "jeux d'esprit," as he called it himself in one of his letters from this period.[13]

Gauss's most original work during these years was devoted to an analysis of the pension fund for the widows of the university professors in Göttingen. Gauss was asked for such a report in 1845 when a gradual increase in the

---

* In Germany, it was—and still is—customary to commemorate this occasion if a doctoral dissertation turns out to be the start of a distinguished and exceptional academic career. Gauss's anniversary was celebrated in Göttingen because Helmstedt, the university from which he had received his Dr. phil., had been closed and no longer existed.

number of widows made it doubtful that the existing level of pensions could be maintained. The statutes, stipulations, benefits, and fees of the insurance had to be reviewed. The system which Gauss investigated received its income from (very modest) membership fees and from various interest-yielding endowments. The size of the pensions fluctuated and was determined according to the financial situation of the insurance. The general subject of Gauss's investigations, the first analysis of the fund, was concerned with the effects of an anticipated major increase in membership, coupled with demands for an increase of the pensions. Gauss used recent mortality tables and the historical information which was available for the insurance. He made long computations, using the available handbooks and as many real data as he could obtain. The surprising conclusion, reached in 1851 after six years' work, was that the system was sound and that its current pensions could even be increased. Gauss further recommended that the membership should be held down; a radical revamping would be necessary if the present growth rate was to be maintained over the next 20 to 30 years.[14]

Gauss's calculation is detailed and careful (though, as usual, not without mistakes in the computations), showing familiarity not only with the mathematical theory of insurance, but also with purely economical aspects such as the assessment of the bonds which the fund owned, or necessary capital investments such as repairs in the buildings it possessed. Again, the task was much bigger than anticipated, but Gauss seems to have enjoyed the work. The correspondence contains quite a few detailed discussions which show his involvement and interest. For the insurance, the analysis was invaluable; Gauss himself must have been gratified that he could, in spite of his fiscal conservatism, suggest a temporary increase of the pensions.*

One of the reasons why Gauss was so involved in his work for the fund was that it gave him an opportunity to apply his practical financial skills and knowledge. In his booklet *Gauss zum Gedächtnis*, Sartorius claims that Gauss could easily have managed the finances of his country; he indeed was, as his estate shows, very skillful in his own affairs and left a considerable fortune, mostly in bonds[15] (riskier but with a better interest rate than today) issued by private companies, often railroads, or by the governments of various states, not only German ones. About his speculations, Gauss does not seem to have consulted any of his friends; in the correspondence, we find only occasional requests for help when certain papers had to be redeemed or interest claimed. Otherwise, we have little knowledge of the details of his transactions—only of their ultimate success. There is one notable exception—a lengthy passage in the correspondence which betrays a considerable, even emotional, involvement. In 1844, Gauss paid a deposit for railroad bonds for a new line in northern Hesse (–Darmstadt). When its bylaws were

---

* Schering, the first editor of *G. W.*, and H. A. Schwarz were among the later Göttingen mathematicians who had to concern themselves with the fund.

published, Gauss's investment lost more than 90% of its value because the government was free to nationalize the railroad at any time. Gauss was very critical of the banks which cooperated, ironically remarking that Christian firms, not Jewish ones, played a leading role in the transaction.[16] Schumacher was the sympathetic and patient recipient of these complaints. We suddenly see Gauss in the unaccustomed role of the independent-minded proud citizen who is willing to speak up—quite a contrast to his attitude during the political crisis of 1837/38 or during the upheavals of the Napoleonic and post-Napoleonic eras, with all their upsetting changes and humiliations.

In the last 15 years of his life, Gauss seems to have settled perfectly into the life of a middle class citizen, free of deep conflicts or commotions. Such a picture is not wrong, but it would indeed be misleading to try to explain this development in terms of Gauss's personal experiences alone. We saw there were several reasons for his gradual withdrawal and the increasing fragmentation of his activities; the price Gauss paid does not necessarily appear too high, and it certainly was not in the eyes of his admiring contemporaries. The years between 1815 and 1848 were not an age for heroes, least of all in Germany. Politically, they were marked by the prevalence of conservatism of a reactionary kind, most conspicuously expressed by the existence and polity of the Holy Alliance; socially, they were characterized, at least in Germany, by the gradual formation of a genuine middle class. This middle class attained, by the middle of the century, a political role, essentially developing the values of classical liberalism.[17] Gauss belonged to this same development, but not to its main stream, and he could never adapt himself to the current progressive political convictions of his class. Against this background, the importance of his financial transactions and his disappointment in the careers of his younger sons become clear. In his social ambitions and dreams, Gauss was very much in tune with the general feeling in Germany, but only in this unpolitical sphere where he could identify with his new class. In 1848, when liberalism made itself politically felt for the first time in Germany, Gauss rejected it.

Again, it was a question of generations, not merely with regard to Gauss's individual age, but also because he had to consider his middle class status as a personal achievement, in marked contrast to the majority of the leaders of the revolt who had a middle class background. We see a constellation similar to Gauss's situation 50 years before, when we tried to understand his attitude towards the French Revolution and the romantic movement in Germany.

Closely related to Gauss's political and social views were his religious beliefs. Despite his strong roots in the Enlightenment, Gauss was not an atheist, rather a deist with very unorthodox convictions, unorthodox even if measured against the very liberal persuasions of the contemporary Protestant church. Again, we do not know many details. There are only scattered remarks in the correspondence and a very curious document—notes which the physiologist R. Wagner made after several "metaphysical" conversations which he had with Gauss in the last two months of 1854, shortly before his death.

Judging from the correspondence, Gauss did not believe in a personal god. An essential part of his credo was his confidence in the harmony and integrity of the grand design of the creation. Mathematics was the key to man's efforts to obtain at least a faint idea of God's plan. Obviously, Gauss's beliefs had a strong resemblance to Leibniz's system, though they were much less systematic and explicit. Typical are Gauss's conviction of the existence of other intelligent life, purely on the basis of statistical conjecture, and the suggestion that man would enjoy a much better geometric intuition after death, allowing him to perceive directly which of the possible geometries was the correct one (presumably and hopefully a hyperbolic, and not the Euclidean geometry). In the correspondence with Gerling, there were some discussions of contemporary mystical experiments, table-turning etc.; these experiments were completely and definitely rejected by Gauss.[18]

Wagner, a conservative and pious man who spent much of his energy in battles against materialism in science, had several interviews with Gauss which he wanted to use for an article after Gauss had died. Because of energetic protests from Gauss's friends and members of the family—Therese simply called Wagner's elaboration *Sudelei* (scribblings)—the article was never published according to Wagner's intentions, and only recently were his original notes found. They contain verbatim transcripts of several passages of the conversations. It is difficult to see how Wagner could hope to use Gauss convincingly for his case, so unconventional were his opinions. But Wagner's good fight was quite desperate, and he may have preferred deism to atheism. The conversations show that Gauss firmly believed in the immortality of the soul and in some sort of life after death, but certainly not in a fashion which could be interpreted as Christian. We quote from a few characteristic passages, starting with Gauss's comments to a list of biblical quotations which he had compiled:

These passages refer to immortality. Right now, I cannot tell you where the compilation comes from. I just cannot find them too convincing or coherent. You seem to believe much more in the bible than I do, and you are much happier than I am. I must say whenever I used to see people of humble station, simple artisans who could believe from the depths of their hearts—this always made me jealous. Tell me, how does one do this? . . . Were you perhaps fortunate enough to have a believing father or mother?[19]

CHAPTER 15

Gauss's Death

There is not much to report from the last five years of Gauss's life. He continued, as well as he could, his observations and correspondence. His two old friends, Schumacher and Lindenau, died in 1850. Gauss had seen Lindenau for the last time during the jubilee the previous year. Of the older friends, only Gerling and the mercurial—"batlike" he called himself— Alexander von Humboldt stayed in touch. Humboldt, considerably older, outlived Gauss and died in 1859 at the age of 90.[1] Observing himself, Gauss complained about his progressive physical and mental deterioration, but there were no major new ailments, only loss of memory and the ability to concentrate and to work. He spent much of the time reading newspapers and light contemporary literature. But even at this late stage of his life, Gauss had some memorable and inspiring experiences, most notably Riemann's address "Über die Hypothesen, welche der Geometrie zu Grunde liegen" in 1854, one year before his death. Riemann gave his lecture in fulfillment of the requirements for attaining the permission to lecture, the *venia legendi*, at the University of Göttingen. Of the three topics which Riemann had to submit, Gauss chose the second, contrary to the custom and to Riemann's expectation that the first theme would be selected. Gauss was clearly interested in the subject, and Weber reports how excited and full of praise Gauss was on the way home from the lecture.*[2]

Gauss must have sensed the importance of Riemann's talk and that it was connected with many facets of his own work of more than fifty years before, but he does not seem to have made any substantive comments. This should not come as a surprise; the admittedly sharp-tongued and critical Jacobi remarked as early as 1849 how uninterested—and presumably unable— Gauss was to participate in concentrated mathematical discussions.[3] Jacobi, one of the few German mathematicians to do so, had come to Göttingen to honor Gauss on the occasion of his jubilee. Gauss's virtually last scientific

---

* Though he spent some time in Göttingen as a student, Riemann cannot be called a direct student of Gauss. For him, the decisive influence came from Berlin, specifically from his teachers Jacobi, Eisenstein, and Dirichlet.

exchange was about a modified Foucault pendulum[4]—again, practical and concrete problems were now, it seems, easier and more accessible than Riemann's abstract and complicated ideas.

In January 1854, Gauss had a thorough physical examination. The diagnosis was dilatation of the heart and did not leave any hope that he might live much longer. Subsequently, his health improved again; it was during this period that Riemann was given the opportunity to deliver his lecture. Gauss also attended the ceremonial opening of the railway link between Göttingen and Hanover. By the beginning of August, his health had deteriorated again and he was no longer able to leave his house; even inside the house, he could not move much because his feet were swollen and barely allowed him to walk. On December 7, the end seemed to have arrived, but Gauss fought it off once more. He died early in the morning of February 23, 1855, after steadily losing strength since mid-January. His watch is reported to have stopped minutes after his death.

Gauss is buried in Göttingen. At the funeral, which was attended by high officials from the government and university, his son-in-law Ewald eulogized Gauss as a unique and incomparable genius. The speech is reprinted in Sartorius' *Gauss zum Gedächtnis*. One of the pall bearers was Richard Dedekind, at that time a 24-year-old student of mathematics. For Dedekind, as for others of his and the previous generation, the memory of Gauss was a lifelong inspiration. In several appendices to lectures of his teacher Dirichlet, Dedekind explained and expanded the last four sections of *Disqu. Arithm.*; the last edition of their *Vorlesungen über Zahlentheorie* appeared in 1893.[5]

Most of the scientists who attended the funeral were not mathematicians; the younger colleagues and pupils who had been close to Gauss were more interested in applied research than in pure mathematics. This, of course, reflects Gauss's predilections in his later years, but also the very practical bent of the times. Among those from the younger generation who attended the funeral were the astronomer Klinkerfues, the geologist Sartorius, and the incomparable Weber.

Gauss's brain, with its, as it turned out, exceptionally deep and numerous convolutions, has been incorporated in the anatomical collection of the University of Göttingen.

Epilogue

(INTERCHAPTER X)

Ultima latet.

Mathematicians are only rarely the subjects of biographies. The genre is more concerned with the so-called historical personalities, politicians, military men, or even artists. Though such biographies will provide some historical background one does not assume that this is really needed. We are supposed to have an intuitive understanding of political, military, or artistic questions and are, above all, able to communicate directly with the author and his subject.

The figure of Gauss is far enough removed from us and close enough to our minds to stir our historical interest and curiosity. It is obvious that any scientist's work, even that of an inspired genius such as Gauss, can only be understood within its contemporary scientific framework; I hope it has been clear that the efforts necessary for an understanding of Gauss's life and work are richly rewarded in contemporary coin. An attentive reader may well argue that from this biography he has not learned sufficiently many details of Gauss's life and work or of the lives and work of his contemporaries and (scientific) predecessors. Though I may have simplified my task to an inordinate degree, it may in fact be easier to depict Gauss's life than that of other, lesser mathematicians and scientists. His exceptional genius has a timeless quality which spans the ages, lighting our way through an otherwise arduous and obscure maze of historical details.

Our views of the past will always contain a strong element of revisionism. Without constant and often unconscious comparisons with our own immediate concerns, our interest in the past would never survive. It is difficult to avoid (ahistoric) misunderstandings but they are often welcome and, to a degree, necessary: there must be some connection with our own concerns, with the problems and demands of our own age. In theory, it may appear—and possibly rightly so—impossible to reconcile the conflicting requirements of historical honesty and justifiable historical involvement, but we are entitled to have our turn at solving this time-honored riddle.

The opposite of the historian, in the sense we use the word, is the antiquarian who strives to reconstruct the details of the past. He lives in

Borges' libraries and broods over maps which correspond exactly to the outside world, even in their scale. One could well write a biography of Gauss on such lines because there are an enormous number of details to be dug up by whoever wishes to perfect his historical miniatures. To collect and narrate these details would be an interesting and entertaining task.

History, as we see it, is selective—a word, alien and odious to the dedicated antiquarian. Though tempting, excessively subjective eclecticism must be avoided as well.

Being basically antiquarian by nature, the current widespread interest in the history of mathematics is hard to understand. Most of the time, history will be neither interesting nor illuminating: the genesis of the majority of mathematical theories is obscure and difficult. Often, today's presentation of a classical topic will be much more accessible and concise than it could ever have been when it was developed. Gauss's ideas, which are the subject of this book, are interesting not because they are old but because Gauss was such a genius as a mathematician and a scientist. Many have survived and are a source of inspiration even today. So this biography, though historical in character, is motivated by our contemporary interest.

It should be clear from what was said above that the history of mathematics uses the same methods, approaches, and arguments as are used in general history. Just as one cannot be a political historian without understanding the political history of the period one considers, one has to have an understanding of the mathematics of the period whose mathematical history one studies. I may well have failed in this because there is so much the biographer has to know. A basic problem which affects this biography or any similar attempt is that a real, serious history of mathematics, especially of the 18th and 19th centuries, has never been written. There exist only scattered investigations of a detailed and specialized nature, but no survey of the evolution of our current mathematical ideas. If such a history existed, it would have to be recent; each generation (of mathematical ideas, not of mathematicians) would have to rewrite it. Among the rudiments of such efforts, Felix Klein's lectures on the development of mathematics in the 19th century are the best specimen but even they are uneven, in part sketchy, and naturally reflect the viewpoint of the first decade of this century if not Klein's personal opinions. In a way, they are now a historical document—the views of an educated and eminent mathematician—and not the required guidebook of the kind described above. Though entertaining and valuable, Klein's lectures sometimes strike one as strange and barely understandable—there is already at least one more layer of historical rubble, curiously intriguing but often irrelevant, reminding the reader of the seven layers that buried ancient Troy.

This biography is but a fragment, a contemporary snapshot, perhaps one small step towards a comprehensive and conclusive *Life of Gauss*—a work that will in all likelihood never be written. Maybe it will provide some material for the unwritten intellectual history of the last 200 years, but it would do so only in the naive sense. Historians may be tempted to use the book as secondary source, but it does not give a definitive historical interpretation. The final *Life of Gauss* would be voluminous. Gauss was an incomparable and timeless genius; his final *Life* might well have to be final indeed—to be written at the end of the history of mathematics.

The Organization of Gauss's Collected Works

Several decades were needed to compile and edit Gauss's collected works; while they were being edited, new material was found, and the editors and their policies changed. The original design, an arrangement by topic, was never dropped though it was not possible to pursue it consistently. The first two volumes are consequently devoted to number theory, the third to analysis, etc. This system was maintained for the first six volumes; Vols. VII–XII contain supplementary material. From the beginning it was clear that the edition could not confine itself to what Gauss himself had published, and Vol. II already contains manuscripts from the *Nachlass*, but the criteria of what to include were gradually expanded during the course of the edition. This change in attitude, together with the fact that new material was discovered and old material better understood, accounts for the rather large number of supplementary volumes. The later volumes contain, among other things, most of the scientifically important passages from Gauss's correspondence. To provide some orientation for the reader, we give here a list of the subtitles of the different volumes; Appendix C consists of a detailed index of the works of Gauss which should help the reader to identify and locate any specific paper.

Vol. I. Disquisitiones Arithmeticae
Vol. II Höhere Arithmetik. Higher Arithmetic
Vol. III. Analysis
Vol. IV. Wahrscheinlichkeitsrechnung und Geometrie. Probability and Geometry
Vol. V. Mathematische Physik. Mathematical Physics
Vol. VI. Astronomische Abhandlungen und Aufsätze. Astronomical Treatises and Papers
Vol. VII. Theoretische Astronomie. Theoretical Astronomy
Vol. VIII. Arithmetik, Analysis, Wahrscheinlichkeitsrechnung, Astronomie. Arithmetic, Analysis, Probability, Astronomy
Vol. IX. Geodäsie. Geodesy
Vol. X,1. Arithmetik, Algebra, Analysis, Geometrie, Tagebuch. Arithmetic, Algebra, Analysis, Geometry, Diary

The original edition of the collected works has been out of print for several years, but a reprint is available, as is most of Gauss's correspondence. In the bibliography, the entries which are in print have been marked by (∗). The various volumes of the correspondence were edited without a general plan or common criteria; unfortunately, only the correspondence with Bessel contains a reliable and useful index. Recently, the Göttingen library published two supplementary volumes, containing new, hitherto unpublished correspondence with Schumacher and Gerling. They are quite substantial, 244 and 124 pages long, and shed new light on some interesting aspects of Gauss's private life, specifically the conflicts with his sons Eugen and Wilhelm, but there are no dramatic revelations. Some of these later published letters were found only recently, others had been deleted by the original editors of the correspondence because they contained supposedly sensitive and private material. Recently, Kurt-R. Biermann republished the correspondence between Gauss and the Humboldt brothers in an excellent new edition. Many of the Humboldt letters are unfortunately lost, but the remaining material is of high scientific and historical interest. Biermann succeeded in compiling a complete and reliable list of the correspondence with Alexander v. Humboldt, including the letters which are lost.

The booklet *Gauss zum Gedächtnis*, published immediately after Gauss's death, is not a genuine primary source, but should be mentioned because it is the only major "eyewitness" account by one of Gauss's immediate contemporaries and friends. It has to be read very critically because its author, the geologist Sartorius [von Waltershausen], did not have a good understanding of Gauss's work and personality. The booklet should be compared to a dusty, faded photograph—but it is a photograph, and not a painting.

There are numerous other sources for additional original material, though of minor importance. Some of them are listed in the bibliography; compare also the remarks in Appendix B. The journal *Mitteilungen der Gauss-Gesellschaft* should be mentioned as the most important modern publication devoted to the cultivation of Gauss's memory. It contains new primary material, particularly previously unpublished letters, essays, and interesting reproductions of little-known pictures of Gauss and his circle.

It should not be expected that the discovery and publication of new documents would lead to important revelations about Gauss's scientific or private life. The last new document of this kind was the diary, which was discovered in 1898. The published sources provide ample material for a broad and thorough understanding of Gauss.

Practically all the original material has been collected and is easily accessible in the archives of the *Niedersächsische Staats- und Universitäts-bibliothek Göttingen*. It is impressive to study the logbooks of Gauss's astronomical observations, to see his enormous computations, or to study his "doodles" and other less conscious expressions of his personality. All this does not add much to what we already know. There is some peripheral material at other places, in the local museum in Brunswick, in the Archives of the former St. Petersburg Academy, and possibly in Berlin, but most of this is of sentimental or antiquarian interest and does not have to be studied if one is not interested in some specific detail.

Gauss published virtually all his important papers in Latin. This was already slightly archaic in his time and makes them inaccessible to most present-day mathematicians. The German summaries which Gauss wrote for many of his shorter papers are helpful; in Appendix C, † denotes the papers for which such summaries exist. Most of the major papers were translated into German in the course of the 19th century, some also into English and French (and other languages). The French edition of *Disqu. Arithm.* appeared as early as 1807, a sign perhaps that Latin was even then less of a lingua franca than is usually assumed.

Today, *Disqu. Arithm.* is available in German and English. The German edition first came out in 1889, under the title *Untersuchungen über höhere Arithmetik*. In addition to *Disqu. Arithm.* the volume contains the translations of Gauss's other number-theoretical papers, to the extent they were published during Gauss's lifetime. The English translation is recent (1966, Yale U.P.) and not everywhere reliable.

Theoria Motus . . . was also translated into German, English and French. The English edition, first published in 1857 and translated and introduced by Rr.-Adm. C. H. Davis was in print until very recently (Dover Publications, Inc.). There are French, German, and English translations of *Disqu. gen.*, Gauss's basic paper in differential geometry, and French, German, and Russian translations of *Intensitas vis magn. terr.* None of these is in print.

Several of Gauss's minor works were published in German in the series *Ostwald's Klassiker*, a collection of "classical" scientific treatises, which was founded and directed by the positivist chemist W. Ostwald before World War I. One of these booklets contains Gauss's various proofs of the law of quadratic reciprocity (*theorema fundamentale*), another his paper on the least squares. The series was revived recently, but the old Gauss booklets have not yet been reprinted. A recent volume contains Gauss's mathematical diary, also to be found in Vol. X of the collected works, together with comprehensive annotations.

A Survey of the Secondary Literature

The scientific literature of the 19th and 20th centuries contains innumerable references to Gauss's work, along with interpretations, commentary, etc. It would be futile to try to discuss them all; even a bibliography can only contain a small selection. From the secondary literature, two collections of essays stand out, because of their quality as well as their accessibility.[1] One is contained in Vols. X,2 and XI,2 of the collected works of Gauss [cf. Appendix A], the other is the centennial volume *Gauss zum Gedächtnis* [Lipsiae 1957, ed. W. Reichardt]. This appendix provides short summaries of the contents of these volumes.

We start with the older essays in Gauss's collected works.

1. Bachmann, Ueber Gauss' zahlentheoretische Arbeiten. Bachmann was a good and knowledgeable number theorist. His essay is reliable and helpful. Nearly two thirds of it is taken up by a careful anaylsis of *Disqu. Arithm.*, the rest is devoted to a comparison of the various proofs of the law of quadratic reciprocity and the higher reciprocity laws. There are also a few pages devoted to Gauss's contributions to analytic number theory.

Bachmann's work is careful, cautious and replete with remarks about the work of later number theorists, but there is comparatively little about Gauss's predecessors. Where both works cover the same ground one would perhaps prefer Dirichlet/Dedekind to Bachmann; the former gives a better idea of the connections between Gauss's work and that of his successors.[2] Still Bachmann's essay is worth studying. It was written for the historically interested mathematician and does not contain any extraneous, specifically biographical, material. Its major shortcoming, for today's reader, is the fact that it was written before World War I, nearly 70 years ago, and does not reflect the research and the results of the last 60 years. Of course, this applies to all the essays in the Collected Works, but the lack of reference to contemporary research is perhaps most disadvantageous in number theory.

2. Schlesinger, Ueber Gauss' Arbeiten zur Funktionentheorie. This essay, first published in 1912, is much more ambitious than Bachmann's. Schlesinger tries to present, as coherently as possible, Gauss's papers in real and complex analysis and to connect them with his other work. The main topics are agM (arithmetico-geometric mean), elliptic functions, the hypergeometric function, and conformal mappings. Schlesinger's main problem was that he had to extrapolate comparatively much from unfinished, fragmentary papers. Despite a number of questionable details and assumptions one can consider the essay as the standard source and reference for Gauss's work in this area. The second major objection which one can make is connected with Schlesinger's extremely ahistorical approach; he reconstructs Gauss's work as if he were a contemporary of Schlesinger, sharing the then current concept of analysis. Markuschewitsch's essay in Reichardt's volume, which will be discussed below, constitutes a partial antidote to Schlesinger though it only differs in details and is actually based on Schlesinger's work. Schlesinger's boldest assumptions are about Gauss's work on ϑ-functions and modular forms, two areas which were at the center of research at the turn of the century. Almost certainly, Schlesinger overestimates Gauss's grasp of the two theories.

Schlesinger's is one of the few essays which contain an extensive and useful index. Because of his very expansive interpretation of his topic, there is a substantial overlap with Bachmann's (number theory) and Stäckel's (geometry) essays.

3. Ostrowski, Ueber den ersten und vierten Gauss'schen Beweis des Fundamentalsatzes der Algebra. This paper is quite different from the other essays. Its subject is the two "analytic" proofs of the fundamental theorem. Ostrowski's main point is to show that Gauss's not quite rigorous proofs can be completed and are acceptable by contemporary standards of "rigor". Gauss's dissertation contains the first genuine proof of the fundamental theorem.

4. Stäckel, Gauss als Geometer. This essay, first published in 1917, is in its approach and motivation similar to Schlesinger's, but less sweeping in its generalizations and conclusions. Stäckel does not proceed chronologically; he starts with the foundations of geometry and finishes with the other area in which Gauss's contributions are particularly noteworthy, differential geometry. The middle chapters deal with Gauss's geometrical interpretations of "nongeometric" mathematical problems and his work in the area of elementary geometry and (spherical) trigonometry. Stäckel's account is very reliable, but he seems to underestimate the rôle of geometric thought in Gauss's work, perhaps misled by Sartorius' statement.[3] Stäckel does not establish connections between the different areas of geometry in which Gauss worked, most notably between differential geometry and the foundations. Probably because of contemporary curiosity, Stäckel spends comparatively much time with explanations of Gauss's ideas about *geometria situs* (topology), an area to which he did not contribute anything systematic. He may have influenced the development of the field through his students Listing, Möbius, and, perhaps, Riemann.

5. Bolza, Gauss und die Variationsrechnung. This thorough account gives an excellent idea of Gauss's contributions to an area which was at the center of research at the time of Gauss. Bolza explains not only Gauss's work, but also its historical place with regard to contemporary, older, and subsequent research. Bolza's main sources are the three papers *Principia generalia, Disqu. generales* (the paper on differential geometry) and *Allg. Lehrsätze* (magnetism). The discussion of *Prin. gen.* gives Bolza a convenient reason for presenting the history of the calculus of variations up to about 1850. Besides Gauss's work, the history of the discovery of Green's theorem and of the δ-symbol are of special interest. Bolza's account of *Disq. gen.* is strictly classical. Because of the changed perspective in differential geometry, Bolza's survey is not quite satisfactory: Gauss's work is of greater contemporary interest than one would assume judging from Bolza's report. The—today—most striking detail in *Allg. Lehrsätze* is Gauss's use of what Riemann called Dirichlet's principle. Bolza closes his discussion with the remark that Gauss's conclusions cannot stand up to the objections which were first formulated by Weierstrass. Though Bolza does not treat an area which is as significant for Gauss's work as number theory, he provides the reader with an excellent and characteristic survey of the field, of Gauss's place in the history of the subject, and of his mathematical thinking.

6. Maennchen, Gauss als Zahlenrechner. This is a systematic essay in which the author tries to give some clues for an understanding of the roots of Gauss's genius. Maennchen is not concerned with Gauss's minor numerical work, but rather with his attitude towards numbers and numerical calculations. Though slightly speculative, Maennchen's conclusions about the interaction between calculations and "theoretical" number theory do not seem too far-fetched. Of *particular* interest are Maennchen's remarks about the individuality that many specific numbers possessed in Gauss's thinking.

7. Geppert, Ueber Gauss' Arbeiten zur Mechanik und Potentialtheorie. This is the last essay in Vol. XI,1 of the collected works and completes the account of Gauss's theoretical work. Geppert treats the mathematical aspects of Gauss's work on the rotation of the Earth, his principle of least constraint, the paper on the attraction of ellipsoids, his contributions to potential theory and mechanics. Geppert is not at all historically oriented and gives very concise accounts which can be read as first introductions. There is much overlap with Bolza's essay; the areas which do not overlap are quite marginal, with the exception of the paper on the principle of least constraint in mechanics.

8. Galle, Ueber die geodätischen Arbeiten von Gauss. This is a valuable survey of Gauss's geodetic work; here, Gauss does not appear as the famous mathematician, but as a most influential and eminent geodesist. In addition to a detailed account of the history of Gauss's survey, the paper contains an exposition of the method of least squares and an appendix about the different types of heliotrope which Gauss built.

9. Schaefer, Ueber Gauss' physikalische Arbeiten (Magnetismus, Elektrodynamik, Optik). This essay gives an account of Gauss's contributions to experimental and theoretical physics; there is considerable biographical, but little historical information. Its theoretical parts overlap with several other essays, particularly Bolza's. The main subjects are magnetism, "galvanism", and electricity theory, including a description of the Gauss–Weber telegraphs (with two illustrations), and optics (theory of telescopes, systems of optical lenses). The detailed and careful exposition of Gauss's fragments on electrodynamics is worth mentioning.

10. Brendel, Ueber die astronomischen Arbeiten von Gauss. This is a good essay with much scientific and biographical information. In addition to the observations, Brendel describes the instruments with which Gauss worked. The second half of the essay is devoted to a presentation of Gauss's work in theoretical astronomy. The most important themes are Gauss's (very early) theory of the Moon, the different stages in Gauss's development from the calculation of the Ceres orbit in 1801 to the methods of *Th. mot.* (1809) and perturbation theory, most importantly the Pallas perturbations.

The Gauss centennial volume of 1957 contains essays of varied direction and interest. We confine ourselves to surveys of those essays which complement the ones in the collected works and contribute to our *contemporary* understanding. The volume also contains some very nice, comparatively little-known illustrations, among them a good, colored map of Gauss's geodetic survey (from Vol. IX of *G.W.*). The first essay we mention is

11. Rieger, Die Zahlentheorie bei Gauss. This is an ideal supplement to Bachmann's more extensive account on which it relies in many of the details. Rieger explains various connections between the work of Gauss and contemporary (i.e., modern) research; he also explains much of what Gauss did in modern terminology. His excellent bibliography has more than 100 entries and includes papers by Hilbert, Minkowski, Tagaka, Deuring, Hasse, and Weil.

12. Kochendörffer, Gauss' algebraische Arbeiten. Perhaps the most useful feature of this brief paper is a section on those algebraic concepts whose germs occur in Gauss's work. Algebra, of course, is the area where Gauss's distaste for unnecessary abstractions is most noticeable and consequential.

13. Blaschke's and *Klingenberg's* papers on Gauss's geometric work are both ahistoric. They describe the developments which were inaugurated by Gauss in this area.[4]

14. Markuschewitsch, Die Atbeiten von C. F. Gauss über Funktionentheorie. This is a useful and welcome complement to Schlesinger's essay, though it does not contribute any new facts. The article is basically historically oriented, and Markuschewitsch is more cautious than Schlesinger in his assumptions about Gauss's fragmentary and only posthumously published work. The essay contains an instructive explanation of Gauss's analytic proofs of the fundamental theorem of algebra.

15. Gnedenko, Ueber die Arbeiten von C. F. Gauss zur Wahrscheinlich-keitsrechnung. This is a useful compilation of Gauss' contributions to probability theory and statistics; it has no counterpart among the essays in the Collected Works. Gnedenko's essay contains an extensive bibliography.

16. Volk's essay covers Gauss's work in astronomy and geodesy. It is concise and useful and particularly strong on the details of the Pallas perturbations.

17. Falkenhagen, Die wesentlichsten Beiträge von C. F Gauss aus der Physik. This article contains a detailed summary of Gauss's work on terrestrial magnetism, including a nice picture of the interior of Gauss's magnetic observatory. Falkenhagen sketches the developments after Gauss and gives a good survey of the theoretical and practical impact of Gauss's work. There are brief summaries of Gauss's other physical works.

The best comprehensive account of Gauss's work is still the one which Felix Klein gave in his *Entwicklung der Mathematik im* 19ten *Jahrhundert.* Its shortcomings are very much those which were pointed out in our discussion of Schlesinger's essay; cf. my remarks in the introduction.

Of the major biographies of Gauss, Dunnington's is by far the most important. Its strength is the enormous amount of material and the book contains; it is indispensable and irreplaceable as a reference. We specifically mention the following outstanding features:

(i) Very fine illustrations with portraits of Gauss, his family, and many of his colleagues and students.

(ii) A (not very enlightening) survey of the fates of Gauss's children and their descendants.

(iii) A listing of the books which Gauss borrowed from the library of his university when he was a student (with the exception of one semester for which the records are missing). The list is very useful and contains scientific books and journals as well as belles lettres.

(iv) A list of courses taught by Gauss at Göttingen between 1808 and 1854.

(v) A complete chronological bibliography of Gauss's works.

(vi) A helpful, but rather indiscriminately compiled bibliography of the secondary literature; cf. my notes in the bibliography of this book.

(vii) Gauss's testament, a curious, but unimportant document.

An Index of Gauss's Works

This appendix consists of a combined title and catchword index for Gauss's collected works. Papers that were published, or at least entitled, by Gauss himself are listed with their titles; smaller fragments are often identified by their topics and not always by the titles which were given to them by the editors of the works. The first and second headings of the tables of contents of the individual volumes should be used as additional reference.

A detailed and comprehensive index of Gauss's works does not exist. There is also no satisfactory index to the correspondence.

Symbols: * published posthumously.
       † with German summary.

Theorematis arithmetici demonstratio nova † II ⟨1808⟩

Theorematis de resolubilitate functionum algebraicarum integrarum in factores reales demonstratio tertia† III ⟨1816⟩

Theorematis fundamentalis in doctrina de residuis quadraticis demonstrationes et amplitiones novae † II ⟨1818⟩

Theoria attractionis corporum sphaeroidicorum ellipticorum homogeneorum V ⟨1813⟩

Theoria combinationis observationum erroribus minimis obnoxiae I & II † IV ⟨1823⟩

Theoria interpolationis methodo nova tractata* III

Theoria motus corporum coelestium in sectionibus conicis Solem ambientium (Summary in Vol. VI) VII ⟨1809⟩

Theoria residuorum biquadraticorum I & II † II ⟨1828⟩ & ⟨1832⟩

Theorie der Bewegung des Mondes* VII

Ternary forms, geometric interpretation* II, VIII

Transcendental trigonometry* X

Über das Wesen und die Definition der Functionen* VIII

Über den d'Angos'schen Cometen* XII

Über den Heliotrop IX ⟨1821⟩

Über die achromatischen Doppelobjective besonders in Rücksicht der vollkommenen Aufhebung der Farbenzerstreuungen V ⟨1817⟩

Über die Anwendung des Magnetometers zur Bestimmung der absoluten Declination V ⟨1841⟩

Über die bei der Landestriangulirung erforderlichen Instrumente* IX

Über die Frequenz von optischen Doppelsternen* XI

Über die Grenzen der geocentrischen Örter der Planeten VI ⟨1804⟩

Über die Kreistheilungsgleichung* X

Über die Reduction von Circummeridianhöhen* XI

Über die Winkel des Dreiecks* VIII

Über ein Mittel, die Beobachtung von Ablenkungen zu erleichtern V ⟨1839⟩

Über ein neues allgemeines Grundgesetz der Mechanik V ⟨1829⟩

Über ein neues Hilfsmittel für die magnetischen Beobachtungen XII ⟨1837⟩

Über ein neues, zunächst zur unmittelbaren Beobachtung der Veränderungen in der Intensität des horizontalen Theils des Erdmagnetismus bestimmten Instruments V ⟨1837⟩

Über eine Aufgabe der sphärischen Astronomie VI ⟨1808⟩

Übertragung der geographischen Lage . . .* IX

Übertragung der Kugel auf die Ebene durch Mercators Projection* IX

Untersuchungen über die transcendenten Functionen, die aus dem Integral $\int dx/\sqrt{1 + x^3}$ ihren Ursprung haben* VIII

Untersuchungen über Gegenstände der höheren Geodäsie I & II† IV ⟨1844⟩, ⟨1847⟩

Variational calculus* XII

Notes*

1.1   There is an extensive "underground literature" consisting of tales from Gauss's childhood that cannot be verified. A comparatively trustworthy source is [Sartorius], who should have heard many of his stories from Gauss himself.

1.2   Appendix E in [Dunnington] contains a family tree, extending over eight generations and starting with the generation of Gauss's father.

1.3   Gauss's paternal family can be traced back to the year 1600. The last name "Gauss" and its versions "Goos" or "Gooss" is quite common in the area to the north of Brunswick. We know less of Gauss's maternal ancestors, though he himself seems to have been more interested in this side of the family and occasionally was in touch with some of his maternal relatives. See [Bord], [Dunnington], and [Haenselmann].

1.4   Much of what we know of Gauss's father is contained in a letter which Gauss wrote in 1810 to Minna Waldeck, later his second wife. See p. 68.

1.5   Gauss's grandfather had started the complicated process by acquiring a very small house within the city limits. After a short time he sold it again and used the proceeds as a downpayment for a bigger house, which he bought with the help of a substantial collateral from the Burgomaster of Brunswick. In 1800, Gauss's father finally paid off the debt when he sold the house for the sum of 1700 talers. He used the money to buy yet another house which was completely his. [After Dunnington.]

1.6   There is apparently only one photo of the house in which Gauss was born. It is reproduced in [Dunnington] and [Reich].

1.7   It is, of course, possible that Gauss remembered his childhood exceptionally clearly. In any event, it seems futile to search for "hard" facts.

1.8   Prussia had the best educational system among the major German states, though it took time to reach reasonable standards.

1.9   The statements in this and the preceding sentence are part of the "Gauss folklore" and probably go back to Gauss himself.

1.10 One of the childhood stories ends with the pronouncement (in the local dialect) "*Dar licht se*" with which the young Gauss handed in his solution of a mathematical assignment.

---

* Italic numbers refer to original quotations of Gauss.

1.11 The two best-known academies are Schulpforta in Saxony and the *Stift* in Tübingen (Württemberg). The former is intimately connected with the development of the Enlightenment in Germany, the latter with that of the romantic movement.

1.12 See note (4)

1.13 More information can be found in any reasonably detailed history of the German literature or in Arno Schmidt's informative essays *Nachrichten von Büchern und Menschen* I.

1.14 In the column "*Neue Entdeckungen*" ("New Discoveries") of the issue of June 1, 1796.

INTERCHAPTER I

I.1 See, e.g., Moritz, *Anton Reiser* [see below].

I.2 Correspondence with Gerling, letters ## 385–387 from 1853.

I.3 The universities in the Catholic German states remained under the supervision of the Church until the end of the 18th century. The situation was hardly different in the states in which orthodox Protestantism prevailed. The concept of state-run and state-supervised primary and secondary schools emerged in the course of the 18th century.

I.4 Berlin 1785ff.

I.5 See K.-R. Biermann, "Beziehungen zwischen C. F. Gauss und F. W. Bessel" in *Mitteilungen der Gauss-Gesellschaft* 3, 1966.

I.6 Goethe's roots go back to *Sturm und Drang*, too, but he later rejected and despised the romantic movement, which presumably was too irrational for his taste.

I.7 This extensive and complicated question will not be discussed explicitly.

CHAPTER 2

2.1 This appears to be part of the Gauss folklore and I have not been able to find an authoritative reference.

2.2 There is a fair amount of secondary literature dealing with the origins and the development of the university of Göttingen. Probably the best general reference is *Geschichte der George-August-Universität* by Götz von Selle 1933. [Du Moulin-Eckardt] and, more importantly, [Smend] (see Bibliography) contain additional information.

2.3 Such a cartoon was among the memorabilia which Bolyai sent Sartorius after Gauss's death. It is reproduced in [Reich].

2.4 See, e.g., Gauss's letter to Olbers of July 24, 1804 (#43).

2.5 Appendix G of [Dunnington] contains a list of books which Gauss borrowed from the university library.

2.6 Included in the Gauss–Bolyai correspondence.

2.7 ... und ich mit dem damals dort [i.e., Göttingen] studierenden Gauss bekannt wurde, mit dem ich noch heute in Freundschaft bin, obgleich weit entfernt mich mit ihm messen zu können. Er war sehr bescheiden und zeigte wenig; nicht drei Tage, wie mit Plato, jahrelang konnte man mit ihm zusammensein, ohne seine Grösse zu erkennen. Schade, dass ich dieses titellose, schweigsame Buch nicht aufzumachen und zu lesen verstand. Ich wusste nicht, wieviel er wusste, und er hielt, nachdem er meine Art sah, viel von mir, ohne zu wissen, wie wenig ich bin. Uns verband die (sich äusserlich nicht zeigende) Leidenschaft für die Mathematik und unsere sittliche Übereinstimmung, so dass wir oft mit einander wandernd, jeder mit den eigenen Gedanken beschäftigt, stundenlang wortlos waren. [From the autobiographical sketch which Bolyai wrote for the Hungarian Academy of Science.]

2.8 Cf., e.g., the Gauss–Bolyai correspondence, p. 153.

2.9 Kästner did not care for Gauss's proof of the constructibility of the 17-gon—he presumably thought this was obvious.

2.10 Gauss-Bolyai correspondence, letter #11, of April 22, 1799.

2.11 Kästner seems to have thought that the axiom of parallels was not independent of the other Euclidean axioms.

INTERCHAPTER II

II.1 Though Gauss was quite interested in keeping track of his own development and the progress of his discoveries he was never systematic about it, and his own comments cannot be taken literally though they are nearly always correct in *some* sense.

II.2 Mainly in Göttingen. There is some material in Brunswick and in Leningrad (St. Petersburg).

II.3 See the correspondence with Bessel; these letters were written in 1811.

II.4 One finds this often quoted statement in Ewald's obituary, which is included in [Sartorius].

CHAPTER 3

3.1 To use indices means to make use of the representation of the cyclic group of the primitive residue classes by a particular generator.

3.2 Gauss's result contains Wilson's theorem as a special case. It is not the first proof of the theorem.

3.3 This is not Gauss's but Legendre's formalism. Gauss never used it in his publications and, though he was certainly familiar with it, never clearly referred to it. When he pointed out, in §76 of *Disqu. Arithm.*, that theorems should follow from notions and not from notations, he did this in connection with his proof of Wilson's theorem and in reference to a remark of Waring. Gauss used Legendre's formalism in several only posthumously published fragments.

3.4 Gauss's other proofs are similarly elementary but not so direct.

3.5 Kronecker called this proof a touchstone (*Prüfstein*) of Gauss's genius.

3.6 The gap in Legendre's proof consists in the assumption, only proved by Dirichlet considerably later, that any arithmetic progression $ax + b$, $(a, b) = 1$, contains an infinite number of primes.

3.7 Legendre's binary quadratic forms were defined as

$$ax^2 + bxy + cy^2.$$

It is computationally advantageous to assume an even coefficient for the term in the middle.

3.8 Gauss calls this function the *determinant*.

3.9   One finds more about this question in the literature, e.g., in [Edwards].

3.10   It would have been difficult for Gauss to confine himself to the theory of binary quadratic forms. Only much later, more than 100 years after the publication of *Disqu. Arithm.*, an elementary proof of the substance of §287, the real object for the disgression into ternary forms, was found.

3.11   In the course of its several editions, Dedekind added more and more appendices to the book and specifically introduced ideal theory into arithmetic. The five chapters of [Dirichlet-Dedekind] cover the following topics: divisibility, congruence, quadratic residues, quadratic forms, and the class number. The four supplements are devoted to the division of the circle, Pell's equation, the composition of forms, and the theory of algebraic integers.

3.12   K. Hensel wrote the following in his introduction to Kronecker's *Vorlesungen über Zahlentheorie*: "... Gauss hat die Arithmetik zum Range einer Wissenschaft erhoben, aber erst Dirichlet gab ihr, wie schon Kronecker mit Recht hervorhob, wirklich eigentliche Methoden, indem er zeigte, dass und wie man ganze Klassen arithmetischer Probleme entweder lösen, oder wenigstens die arithmetische Schwierigkeit auf eine analytische reduzieren kann. Die Methoden Dirichlet's beruhen wesentlich auf der Einführung des Grenzbegriffes in die Arithmetik ..."

3.13   ... Was da ... [*Disqu. Arithm.* §356] ... steht ist streng dort bewiesen, aber was fehlt, nämlich die Bestimmung des Wurzelzeichens, ist es gerade, was mich immer gequält hat. Dieser Mangel hat mir alles Übrige, was ich fand, verleidet, und seit vier Jahren wird selten eine Woche hingegangen sein, wo ich nicht einen oder den anderen vergeblichen Versuch, diesen Knoten zu lösen, gemacht hätte besonders lebhaft nun auch wieder in der letzten Zeit. Aber alles Brüten, alles Suchen ist umsonst gewesen, traurig habe ich jedesmal wieder die Feder niederlegen müssen. Endlich vor ein paar Tagen ist es mir gelungen—aber nicht meinem mühsamen Suchen sondern bloss durch die Gnade Gottes möchte ich sagen. Wie der Blitz einschlägt, hat sich das Räthsel gelöst; ich selbst wäre nicht im Stande, den leitenden Faden zwischen dem, was ich vorher wusste, dem, womit ich die letzten Versuche gemacht hatte—und dem, wodurch es gelang nachzuweisen ....

3.14   See, e.g., the letters to Bessel of June 28, 1820 (#45), and of March 12, 1826 (#58), or the letter to Dirichlet of Nov. 2, 1838.

3.15   The second volume of Kronecker's collected papers contains a historical review of the law of quadratic reciprocity. One finds relatively complimentary remarks about Legendre's incomplete proof in Gauss's appendix of *Disqu. Arithm.* and in his "Theorematis arithmetici demonstratio nova" of 1808.

CHAPTER 4

4.1   Letters #I to Bolyai, of Sept. 29, 1797, and #III, of Sept. 30, 1798.

4.2   See, e.g., [Dunnington].

4.3   Von meinem Herzog habe ich Ursache zu hoffen, dass er seine Unterstützung auch in der Folge noch fortsetzen werde, bis ich eine bestimmte Lage erhalte. Eine gewisse lucrative habe ich verfehlt. Es hält sich hier ein russischer Gesandter auf dessen zwei junge, sehr geistreiche Töchter ich in der Mathematik und Astronomie hatte unterrichten sollen. Weil ich aber zu lange ausblieb, so hat ein französischer Emigrant das Geschäft schon übernommen.

4.4 Letter #V to Bolyai, of Nov. 29, 1798.

4.5 In a surprising passage, [Sartorius] explicitly refutes the claim, apparently made in a biography of Pfaff, that Pfaff could have had any role in Gauss's dissertation.

4.6 Der Titel gibt ganz bestimmt die Hauptabsicht der Schrift an, indessen ist zu dieser nur ungefähr der 3te Theil des Ganzen verbraucht, das übrige enthält vornehmlich Geschichte und Kritik der Arbeiten anderer Mathematiker (namentlich d'Alembert, Bougainville, Euler, de Foncenex, Lagrange und die Compendienschreiber—welche letztere aber wol eben nicht sehr zufrieden sein werden) über denselben Gegenstand nebst mancherlei Bemerkungen über die Seichtigkeit die in unserer heutigen Mathematik so herrschend ist. [Letter #XVIII to Bolyai, of Dec. 16, 1799.]

4.7 The Norwegian mathematician Wessel and the Swiss scientist d'Argand independently discovered the geometric visualization of the complex domain.

4.8 Judging from the correspondence (they never met), Gauss had a particularly good rapport with N. von Fuss, the Secretary of the Academy, who negotiated for the Russian side.

4.9 Gauss obtained some funds to acquire several good instruments, but there were never any concrete plans to build a regular observatory. See also Gauss's letter #XXIII to Bolyai, of June 20, 1803.

4.10 . . . Sie kennen, liebster Freund, obgleich Mathematik und Astronomie nicht Ihr Fach ist, den grossen Ruhm, den sich Dr. Gauss in Braunschweig erworben hat. Dieser Ruhm ist vollkommen verdient, und der junge Mann von 25 Jahren geht schon allen seinen mathematischen Zeitgenossen vor. Ich glaube, dies einigermassen beurtheilen zu können, da ich nicht nur seine Schriften gelesen habe, sondern auch seit dem Anfange dieses Jahres mit ihm in vertrautestem Briefwechsel stehe. Seine Kenntnisse, seine ausserordentliche Geschicklichkeit im analytischen und astronomischen Calcul, seine unermüdliche Thätigkeit und Arbeitsamkeit, sein ganz unvergleichbares Genie haben meine höchste Bewunderung erregt, und immer vermehrt, je mehr er mir in dem Laufe unsers Briefwechsels von seinen Ideen mittheilte. Dabey liebt er die Sternkunde, vorzüglich die practische Sternkunde enthusiastisch, so wenig er auch aus Mangel an Instrumenten bisher Gelegenheit gehabt hat, letztere zu treiben. Für eine mathematische Lehrstelle hat er eine ganz entschiedene Abneigung: sein Lieblingswunsch ist, Astronom bei irgend einer Sternwarte zu werden, um seine ganze Zeit zwischen Beobachtungen und seinen tiefsinnigen Untersuchungen zur Erweiterung der Wissenschaft theilen zu können . . . [Letter to Heeren of Nov. 3, 1802.]

CHAPTER 5

5.1 Johanna's family seems to have been socially superior to her husband's. Olbers knew her as a girl.

5.2 Gestern Mittag, Freitags, bin ich hier nach einer sehr beschwerlichen Reise angelangt: dir liebes Hannchen, gehört meine erste Feder. Das abscheuliche Wetter, dem ich von Mittwoch Abends um 9 bis Donnerstag Morgens um 12 beständig ausgesetzt war, der durch Chenille, Schlafrock, zwei Kleider und Hemd endlich doch bis auf die Haut durchdringende Regen haben mir diese Art Reisen sehr verleidet: glücklicherweise hat es mir aber doch gar nichts geschadet, und ich bin mit der vorübergehenden Ungemächlichlichkeit davon gekommen.

Schwierigkeiten andrer Art habe ich gar keine gehabt: sicher sind die Wege hier völlig, und nach meinem Passe ist nirgends einmal gefragt. Olbers habe ich nicht ganz wohl getroffen, er hat die Rose auf einer Backe, und darf sich jetzt nicht gut der Luft aussetzen übrigens befindet er und seine Familie sich wohl, alle grüssen dich herzlich. Bis jetzt (Vormittags 9 Uhr) habe ich noch keine neue Bekanntschaft gemacht als die von Dr Focke und dessen kleinen Wilhelm, ein allerliebstes sehr gesundes Kind von zwei Jahren, mit 10 Monat hat er schon ganz sicher gelaufen. Bessel hoffe ich noch heute zu sehen.

Wegen der Angelegenheit in Göttingen räth mir Olbers sehr zu, auf die Sicherheit des Engagements glaubt er könne ich mich völlig verlassen. Sollte aber auch Ernst mit der Sache werden so würde ich es doch wahrscheinlich so einzurichten suchen, dass ich nicht um Michaelis, sondern erst im Laufe des Herbst oder um Neujahr anträte, eben weil ich im nächsten Winter mich auf Vorlesungen nicht wohl würde einlassen können. Mit unsrer Wohnung bleibt es also auf alle Fälle noch beim Alten. Umsonst liebes Hannchen habe ich heute einen Brief von Dir erwartet: ich hoffe, dass nichts widriges vorgefallen ist und ihr alle wohl seid. Schreib mir ja bald wie es dir geht, ob deine gute Mutter ihre Rose wieder los ist (mit Olbers ist es auf der Besserung, ohne dass er etwas braucht, als sich zu schonen) was unser süsser Joseph macht, wie du mit seiner neuen Wärterin zufrieden bist. Hast du Hrn Mengen seinen Wein bezahlt, vielleicht hast du die Summe nicht mehr gewusst, ich meine es war $1\mathcal{T}$ 14 ggr. $9\,d$.

5.3   Ein wunderschönes Madonnengesicht, ein Spiegel des Seelenfriedens und der Gesundheit, zärtliche, etwas schwärmerische Augen, ein tadelloser Wuchs, das ist etwas, ein heller Verstand und eine gebildete Sprache, das ist auch etwas, aber nun eine stille, heitre, bescheidene, keusche Engelsseele, die keinem Wesen wehe thun kann, die ist das beste.

5.4   See Gauss's correspondence with Bessel, specifically Bessel's letter #34 of Oct. 19, 1810, and subsequent letters.

5.5   Vol. III contains "Determinatio attractionis . . .", Vol. VI the calculation of the elements of Pallas for the years 1803–1805, 1807–1809, and Vol. VII additional extensive fragments concerning the Pallas perturbations.

One of Gauss's most interesting discoveries concerns the fact that the quotient between the mean movements of Jupiter and Pallas is rational and exactly 7:18. It shows that Jupiter exerts an effect on the movement of Pallas analogous to the Earth's effect on the movement of the Moon. Gauss mentioned his discovery in his letter to Bessel of May 5, 1812; there is some additional material in Vol. VII, posthumously published from Gauss's notes. The effect is called *libration*.

5.6   Gauss did some practical work with Zach in the summer of 1803.

5.7   Der Spiegel des 10 f. Teleskops, den ich nun zurückerhalten habe, scheint jetzt recht gut geworden zu sein; viel Proben damit anzustellen verhindert der Platz; auch habe ich zum genauen Centrieren jetzt keine Zeit und Lust. Es ist leicht möglich, dass dies Instrument nach meiner Abreise hier in sehr schlechte Hände kommt.

5.8   Letter #306 to Schumacher, of July 6, 1840.

5.9   Neulich habe ich die Freude gehabt, einen Brief von einem jungen Geometer aus Paris LeBlanc zu erhalten, der sich mit Enthusiasmus mit der höheren Mathematik vertraut macht, und mir Proben gegeben hat, dass er in meine *Disquis. Arith.* tief eingedrungen ist . . .

5.10 Meine *Disqu. Arith.* haben mir unlängst eine grosse Überraschung veranlasst. Habe ich Ihnen nicht schon einigemale von einem Pariser Korrespondenten LeBlanc geschrieben, der mir Proben gegeben hat, dass er sich alle Untersuchungen dieses Werkes auf das Vollkommenste zu eigen gemacht hat? Dieser LeBlanc hat sich neulich mir näher zu erkennen gegeben. Dass LeBlanc ein bloss fingierter Name eines jungen Frauenzimmers Sophie Germain ist, wundert Sie gewiss ebenso sehr als mich.

5.11 One of the proofs that follow from the theory of biquadratic residues.

5.12 See, e.g., [Edwards].

5.13 Letter #XXIX to Bolyai, of May 20, 1808.

5.14 There were occasional tensions between Gauss and Harding, probably because it was never made clear what Harding's position was when Gauss came to Göttingen. At least initially, Harding considered himself more independent than Gauss was willing to permit.

INTERCHAPTER IV

IV.1 In the Seven Years War.

IV.2 The Prussian army expected victory but it was disorganized and without a clear chain of command. The old Duke was not capable of bearing the responsibility of supreme command which had been forced upon him.

IV.3 At the time, Gauss's address was Steinweg 22, very close to the gate.

IV.4 Napoleon's brother Jerôme was the ruler of the newly created Kingdom of Westphalia, a buffer state between France and Prussia. Generally, Jerôme is not considered to have been a capable or efficient ruler, but this judgment may well be unfair. Jerôme ruled for less than 10 years.

IV.5 The question of the oath was to attain considerable importance in 1837, during the protest of the Göttingen Seven. Ordinarily, every German civil servant, including teachers and professors, had to give an oath of allegiance to his prince or, later, the constitution of his state.

IV.6 Several times, the question came up in the correspondence with Gerling. Joseph left the service early to join one of the new railroad companies, where he was quite successful.

IV.7 Westphalia imitated France where Napoleon created numerous counts, dukes, and princes.

IV.8 G. E. Lessing is the most important German writer and poet of the 18th century. He lived in Hamburg before accepting the offer to come to Wolfenbüttel.

IV.9 It strongly affected the nobility, which lost most of its political influence.

IV.10 To be a lawyer was a desirable objective for the grandson of an unskilled laborer. This seems to be a safe assumption though we do not know any specifics.

IV.11 See, e.g., Golo Mann, *Deutsche Geschichte des 19. und 20. Jahrhunderts.* Frankfurt 1958.

IV.12 One of the first proponents of a German national literature was Martin Optiz, who lived and worked in the 17th century. Later, the rejection of foreign influences developed into a recurring theme, culminating in the (nationalistic) *Sturm und Drang* and romantic movements. The classical poetry of Wieland, Herder, Schiller, or Goethe was internationally minded.

IV.13   See Hölderlin's poem *Empedokles—An die Deutschen.*
IV.14   Uhland, *Das alte gute Recht* (in *Vaterländische Gesänge*).

CHAPTER 6

6.1   See Johanna Gauss's letter of Nov. 21, 1807, to her friend Köppe. It is reprinted in [Mack].

6.2   The word *Biedermaier*, of undetermined origin, is used to characterize the first decades of the 19th century in Germany. One easily associates a meaning with it, for *Maier* is a common last name in Germany, *bieder* means straight and quiet.

6.3   ... Herzlich leid mein süsser Liebling thut es mir, das mein Schweigen Dich unruhig gemacht hat, alles in unsern Hause gieng den gewöhnlichen Gang; ich befand mich ausser der Sehnsucht nach Dir sehr wohl, der Josepf entbehrte nichts, sondern war sehr Lustig, seine neue Wärterin, wie ich Dir schon gemeldet habe, kam am Freitag an, es ist eine sehr rechtliche stille Person, zwar eine alte Jungfer, aber so kinderlieb, das der Josepf schon am ersten Tage heymisch bey ihr war, jetzt ist er völlig so gern bey ihr als bey seiner Mutter, dies dünkt mir ist mir Bürge, das ich ihn ihr unbedingt anvertrauen kann. er geht täglich spazieren und besucht mit ihr entweder unsere Verwanten oder seine Vorgänger, deren eine ziemliche Menge sind, dies behagt ihn so sehr, das er es durch zeigen auf die Thür und durch ampeln nach derselben sehr deutlich macht, wenn seine Lust im Zimmer zu bleiben aus ist, seine Lebhaftigkeit hat sehr zugenommen, jeder freut sich über ihn und will es nach der Ebeling (der Wärterin) ihrer Versicherung nicht glauben, das dies zarte feine Gesicht einem Knaben gehöre, am 26. ist ohne alle Umstande der 7te Zahn angekommen, dafür aber hat der arme Schelm sein grösstes Gut am Sontage eingebüsst, ich bin, so oft ich daran dencke, unbeschreiblich traurig, doch war ich zu diesem schnellen Entschlusse genötigt, da die Ebelingen nur auf unbestimmte Zeit bleiben kann ... meine Unentschlossenheit, wann ich es thun wolle, war vorzüglich Schuld, das ich Dir am Freitage nicht schrieb, auch glaubte ich, das Briefe an Dich kommen würden, welches aber erst ungewöhnlich spät nach $\frac{1}{2}$10 gescha, die möglichkeit, das Harding früher als Du eintreffen könne, bewog mich Dir den Brief zu schicken, vergieb meiner Eile das kunfuse Cuvert. [Letter of June 30, 1807.]

6.4   Alle Stunden mein theuerstes Hannchen, wo ich keine besondere Beschäftigung habe, weiss ich nicht angenehmer anzuwenden, als wenn ich mich mit Dir unterhalte, wenn ich gerade nichts von Bedeutung zu melden habe. Ich fahre also fort dir zu erzählen, wie ich bisher meine Zeit in Bremen zugebracht habe ... [Letter of July 1, 1807.]

6.5   Dein lieber Brief vom 30., welchen ich soeben erhalte, macht mir ausnehmend viel Freude. Es ist ein unschätzbares Glück, dass unser süsser Joseph den kritischen Zeitpunkt in so guten und zuverlässigen Händen abwarten kann; wenn Du diesen Brief erhältst, ist vermuthlich das schlimmste schon vorbei. Studiert er die Lehre vom Gleichgewicht und von der Bewegung noch fleissig? Die Beschwerden meiner Reise haben meiner Gesundheit nichts geschadet, aber die so sehr veränderte Diät (ich esse hier zuverlässig viermal so viel als zu Hause und doch beschwert man sich noch über meinen wenigen Appetit) hat anfangs einige Obstructionen zugezogen, denen aber duch ein Digestivpulver bald abgeholfen wurde, jetzt fange ich an der epicureischen Lebensart gewohnt zu werden. Olbers glaubt nicht, dass gegen meine Magenschwäche, Blähungen und Obstructionen die Apotheke etwas gründliches vermöge, eher der Keller. Die Diät und Lebensart müssten bei dieser Art von Übel das Beste thun. Unsere gewöhnliche Sorte Rothwein, die Tavelle, hält er für ungesund und glaubt, dass er, wenn auch nicht mein öftres Herzklopfen allein hervorbringen, es doch sehr befördern könne. Sehr gut für den Magen sei zuweilen ein Glas Madera, zur Beförderung der Öffnung empfielt er mir eine Pfeife täglich früh zum Kaffee, übrigens Bewegung u.s.w. Sehr gut würde mir aber vorzüglich zuweilen ein laues Bad sein; dem von Zeit zu Zeit wiederhohlten Gebrauche einer Brunnencur schreibt er die heilsamsten Wirkungen zu; wer weiss, ob wir nicht übers Jahr einander in Rehburg ein Rendesvous geben können. Sehr grosse Lust hat Olbers auch, dereinst einmal mit mir eine Reise nach Paris zu machen, da wir alle beide das französische Theater u. dergl. Narrenpossen eben nicht zu schätzen wissen, so würden wir beide in ein paar Wochen das meiste uns sehenswürdige abthun und in etwa 5 Wochen die ganze Reise vollenden können.

Die Nachricht von dem Waffenstillstande zwischen den Franzosen und Russen scheint sich zu bestätigen und auf einen naher Frieden zu deuten, unterdess sind nun aber die Engländer in Schwedischpommern gelandet. Es sind tolle Zeiten.

Es wird hohe Zeit mich anzukleiden: ich muss also eilig schliessen. Viele Grüsse an deine gute Mutter, sowie an alle unsre Freunde, ebenso als hätte ich sie namentlich und einzeln genannt.

6.6   Albrecht von Haller (1708–1777) is remembered in the history of German literature for his epic poem *Die Alpen* (*The Alps*) which expresses a new appreciation of nature that influenced the *Sturm und Drang* and romantic movements. Before and after his Göttingen years, von Haller lived in his native Switzerland, the home of several major German-language poets of this period.

6.7   For more information, see [Smend]. His article describes how the academy was established, 14 years after the university was founded, as a special institute for the intellectual exchange of its illustrious members.

6.8   In Germany, Metternich's philosophy was pervasive. Things changed only after 1848 when Metternich has to resign and Louis Philippe was deposed.

6.9   See Goethe, *Faust*, der Tragödie erster Theil.

6.10   He ended his career as *Regierungspräsident* in Detmold. This position was comparable to that of a prefect in France.

6.11   Rudolf Borchardt gives an interesting account of these developments in his essay *Deutsche Denkreden* in *Prosa* III, Stuttgart 1960.

6.12   Süssmilch was heavily influenced by earlier English work. There was traditionally more interest in empirical statistical results in England than on the Continent.

6.13   Es ist dem Menschen nichts angenehmers, als die Gewissheit der Erkenntnis, und wer einmal dieselbe geschmeckt, der bekommt einen Eckel für allem, wo er nichts als Ungewissheit siehet. Aus dieser Ursache ist es kommen, dass die Mathematici, welche beständig mit gewisser Erkenntnis umgegangen, einen Eckel vor der Philosophie und andern Dingen bekommen, und nichts angenehmers gefunden, als dass sie ihre genutzte Zeit mit Linien und Buchstaben zubringen können.

6.14   Goethe, who rejected Newton's theory for philosophical reasons, developed a complicated original theory of color. Gauss was never very interested in Goethe's poetry (though he liked it) nor was he impressed by his scientific dilettantism.

The two men never met but we know that Goethe admired Gauss and had his autograph in his collection.

CHAPTER 7

7.1   ... Glücklich fliessen die Tage in dem einförmigen Gange des häuslichen Lebens hin: wenn das Mädchen einen neuen Zahn kriegt, oder der Junge ein paar neue Wörter gelernt hat, so ist das fast ebenso wichtig, als wenn ein neuer Stern oder eine neue Wahrheit entdeckt ist ... (Letter #XXX to Bolyai, of Sept. 2, 1808.)

7.2   Sie luden mich so freundlich ein, Sie zu besuchen, wenn meine Frau sich wohl befände. Jetzt befindet sie sich wohl. Gestern Abend um 8 Uhr habe ich ihr die Engelsaugen, in denen ich seit 5 Jahren einen Himmel fand, zugedrückt. Der Himmel gebe mir Kraft, diesen Schlag zu tragen. Erlauben Sie mir jetzt, theurer Olbers, bei Ihnen ein paar Wochen in den Armen der Freundschaft Kräfte für das Leben zu sammeln, das jetzt nur noch als meinen drei unmündigen Kindern gehörend Werth hat. Erlaubt es der Arzt, so komme ich vielleicht diesem Briefe schon in ein paar Tagen nach. (Letter #101 to Olbers, of Oct. 12, 1809.)

7.3   It was among the family papers that had come down to Gauss's grandson Carl August Gauss. Though it is certainly touching it is difficult to tell whether it reflects a direct emotional experience or is a conventional expression of mourning. We must keep in mind that the times were sentimental. See, e.g., Jung-Stilling's *Lebensgeschichte* [Vol. I., 1777].

7.4   Siehst Du geliebter Schatten meine Thränen? Du kanntest ja, so lange ich dich die meine nannte, keinen Schmerz, als den meinigen, und brauchtest zu Deinem Glücke Nichts, als nur mich froh zu sehen! Selige Tage! Ich armer Thor konnte ein solches Glück für ewig halten, konnte wähnen, Du einst verkörperter und jertzt wieder neu verklärter Engel seyst bestimmt, mein ganzes Leben hindurch alle die kleinlichen Bürden des Lebens mir tragen zu helfen? Womit hatte ich denn dich verdient? Du bedurftest nicht des Erdenlebens, um besser zu werden. Du tratst nur ein ins Leben, um uns vorzuleuchten. Ach ich war der Glückliche, dessen dunkle Pfade der Unerforschliche von deiner Gegenwart, von deiner Liebe, von deiner zärtlichsten und reinsten Liebe erhellen liess. Durfte ich dich für meines Gleichen halten? Theures Wesen, du wustest selbst nicht, wie einzig du warst. Mit der Sanftmuth eines Engels ertrugst du meine Fehler. O wenn es den Seligen vergönnt ist noch unsichtbar uns armen im Lebensdunkel irrenden nahe zu seyn, verlass mich nicht. Kann deine Liebe vergänglich seyn? Kannst Du sie dem armen, dessen Höchstes Gut sie war entziehen? O du beste, bleib meinem Geiste nah Lass deine selige Seelenruhe, die dir den Abschied von deinen Lieben tragen half, sich mir mittheilen; hilf mir, deiner immer würdiger zu seyn! Ach was kann den theuren Pfändern unsrer Liebe dich, deine mütterlichen Sorge, was dein Vorbild ersetzen, wenn du mich nicht stärkst und veredelst, für sie zu leben, und in meinem Schmerze nicht zu versinken!

   25 Okt. Einsam schleiche ich unter den fröhlichen Menschen, die mich hier umgeben. Machen sie mich meinen Schmerz auf Augenblicke vergessen, so kommt er nachher mit verdoppelter Stärke zurück. Ich tauge nicht unter eure frohe Gesichter. Ich könnte hart gegen euch werden, was ihr nicht verdient. Selbst der heitere Himmel macht mich nur trauriger. Jetzt hättest du theure nun dein Lager verlassen, jetzt wandeltest du an meinem Arme unsern Liebling an

der Hand und freutest dich deiner Genesung und unsers Glücks, das wir jeder im Spiegel der Augen des andern läsen. Wir träumten von einer schönen Zukunft. Ein neidischer Dämon—nein kein neidischer Dämon, der Unerforschliche hat es nicht gewollt. Du Seelige schauest nun schon die dunkeln Zwecke, die durch die Zertrümmerung meines Glücks erreicht werden sollen, in Klarheit an. Ist es dir denn nicht vergönnt dem Verlassenen einige Tropfen Trost und Resignation ins Herz zu flössen? Du warst ja schon im Leben so überreich an beiden. Du hattest mich so lieb. Du wolltest so gern bei mir bleiben. Ich sollte mich doch nicht zu sehr dem Gram überlassen, waren beinahe deine letzten Worte. Ach wie fange ich es an ihm zu entgehen. Ach erbitte dir von dem Ewigen—könnte er dir alles abschlagen?—nur das Einzige, dass deine unendliche Seelengute mir stets recht lebendig vorschwebe, damit ich, so gut ich armer Erdensohn kann, dir nachstrebe.

7.5   Gauss alludes to an earlier engagement in his first letter to Miss Waldeck.

7.6   The age was romantically inclined, and it would be surprising had Minna Waldeck not expected some vigorous exertions from her future husband.

7.7   ... Mit klopfendem Herzen schreibe ich Ihnen diesen Brief, von dem das Glück meines Lebens abhängt. Wenn Sie ihn empfangen, sind Sie schon bekannt mit meinen Wünschen. Wie werden Sie, Beste, sie aufnehmen? Werde ich Ihnen nicht in einem nachtheiligen Lichte erscheinen, dass ich, noch kein halbes Jahr nach dem Verluste einer so geliebten Gattinn, schon an eine neue Verbindung denke? Werden Sie mich deshalb für leichtsinnig oder noch schlimmer halten?

Ich hoffe, Sie werden es nicht. Wie könnte ich auch den Muth haben, Ihr Herz zu suchen, wenn ich mir nicht schmeichelte, in Ihrer Meinung so gut zu stehen, dass Sie mich keiner Motive fähig halten könnten, für die ich erröten müsste.

Ich ehre Sie viel zu sehr, um es Ihnen verschweigen zu wollen, dass ich Ihnen nur ein getheiltes Herz anzubieten habe, in welchem das Bild des verklärten Schattens nie erlöschen wird. Aber wenn Sie wüssten, Sie Gute, wie sehr die Verewigte Sie liebte und achtete, Sie würden mich ganz verstehen, dass ich in diesem wichtigen Augenblicke, wo ich Sie frage, ob Sie sich entschliessen können den von der Verewigten verlassenen Platz anzunehmen, diese lebendig vor mir sehe, freudig meinen Wünschen zulächelnd und mir und unsern Kindern Heil und Segen wünschend.

Aber, Theuerste, ich will Sie nicht bestechen bei der ernstesten Angelegenheit Ihres Lebens. Dass eine Selige mit inniger Freude auf die Erfüllung meiner Wunsche herabsehen würde; dass Ihre Mutter, die ich damit bekannt gemacht habe (sie selbst wird Ihnen sagen was mich dazu vermocht hat)—dass Ihr Vater, welcher durch Ihre Mutter darum weiss, meine Absichten billigen und unser aller Glück davon hoffen; dass ich, dem Sie theuer waren vom ersten Augenblicke an wo ich Sie kennen lernte, überglücklich dadurch werden würde, diess alles erwähne ich bloss darum, um Sie zu bitten, um Sie zu beschwören, darauf keine Rücksicht zu nehmen, sondern bloss Ihr eignes Glück und Ihr eignes Herz zu Rathe zu ziehen. Sie verdienen ein ganz reines Glück und müssen sich durchaus durch keine Nebenrücksichten, die ausserhalb meiner Persönlichkeit liegen, von welcher Art sie auch sein mögen, leiten lassen. Lassen Sie mich Ihnen auch ganz offen gestehen, dass, so bescheiden und genügsam ich sonst in meinen Ansprüchen an das Leben bin, es in dem engsten häuslichen Verhältnisse keinen Mittelzustand für mich geben kann, und dass ich da entweder höchst glücklich oder sehr unglücklich seyn muss: und glücklich würde mich selbst die Verbindung mit Ihnen nicht machen, wenn Sie es dadurch nicht ganz würden . . .

7.8   The letter of April 15 is one of our major sources for whatever factual information we have about Gauss's parents and the way Gauss saw them.

7.9   Doch Ein Wort noch: der Grund, warum ich nicht an meine Mutter geschrieben habe, ist weil ich sie gerne überraschen möchte; aber der Grund, warum ich Sie nicht habe schreiben lassen wollen, ist—weil meine Mutter Geschriebnes nicht lesen kann, und Sie es doch nicht wünschen würden, Ihre schöne Seele vor Personen, denen es nicht bestimmt war, ganz gezeigt zu haben.

7.10   Lilienthal, near Bremen, was an important (privately owned) observatory, after Seeberg probably the second in Germany. Its owner, Schröder, was a wealthy lawyer with a strong interest in astronomy.

7.11   Felix Klein was very interested in this question and discussed it with an anti-18th-century slant.

7.12   See, as examples, Gauss's letters #24 of April 1816, #348 of Dec. 11, 1842, or #370 of Sept. 20, 1843.

7.13   This fact may have been the reason for Kronecker's objections who looked at Gauss's work from a different perspective.

7.14 . . . Sie sind ganz im Irrthum wenn Sie glauben, dass ich darunter nur die letzte Politur in Beziehung auf Sprache und Eleganz der Darstellung verstehe. Diese kosten vergleichsweise nur unbedeutenden Zeitaufwand; was ich meine, ist die *innere* Vollkommenheit. In manchen meiner Arbeiten sind solche Incidenzpunkte, die mich jahrelanges Nachdenken gekostet haben und deren in kleinem Raum concentrirter Darstellung nachher niemand die Schwierigkeit anmerkt, die erst überwunden werden muß.

7.15   [Sartorius], p. 82. See also letter #123 to Schumacher of February 12, 1826.

INTERCHAPTER VI

VI.1   *G.W.*, Vol. XII, pp. 57ff.

VI.2   See letter #27 to Bessel, of Nov. 13, 1814, or letter #98 to Olbers, of Dec. 3, 1808. Gauss changed his opinion in the case of Lindenau. Initially, he did not take him seriously, but accepted him later as friend and colleague. (See Letter #61 to Olbers, of July 2, 1805.)

VI.3   The journal's exact name was *Commentationes societatis regiae scientiarum Gottingensis recentiores.*

VI.4 . . . Da in dem angeführten Werke die Untersuchung so weit bereits geführt, und nur die Bestimmung des Zeichens für irgend einen Werth von k noch übrig war: so hätte man glauben sollen, dass nach Beseitigung der Hauptsache diese nähere Bestimmung sich leicht wüde ergänzen lassen, um so mehr, da die Induction dafür sogleich ein äusserst einfaches Resultat gibt: für $k = 1$, oder für alle Werthe, welche quadratische Reste von n sind, muss nemlich die Wurzelgrösse in obigen Formeln durchaus positiv genommen werden. Allein bei der Aufsuchung des Beweises dieser Bemerkung treffen wir auf ganz unerwartete Schwierigkeiten, und dasjenige Verfahren, welches so genugthuend zu der Bestimmung des absoluten Werths jener Reihen führte, wird durchaus unzureichend befunden, wenn as die vollständige Bestimmung der Zeichen gilt. Den *metaphysischen* Grund dieses Phänomens (um den bei den Französischen Geometern üblichen

Ausdruck zu gebrauchen) hat man in dem Umstande zu suchen, dass die Analyse bei der Theilung des Kreises zwischen den Bögen ω, 2ω, 3ω ... $(n-1)\omega$ keinen Unterschied macht, sondern alle auf gleiche Art umfasst; und da hiedurch die Untersuchung ein neues Interesse erhält: so fand Hr. Prof. Gauss hierin gleichsam eine Aufforderung, nichts unversucht zu lassen, um die Schwierigkeit zu besiegen. Erst nach vielen und mannigfaltigen vergeblichen Versuchen ist ihm dieses auf einem auch an sich selbst merkwürdigen Wege gelungen ... (*G.W.* II. p. 156)

VI.5   The relevant reviews are in Vol. IV of *G.W.*

VI.6   See Vol. XII of *G.W.*

VI.7   *Untersuchungen über die Eigenschaften der positiven ternären quadratischen Formen*; Gauss's review is in Vol. II of *G.W*

VI.8   *Disqu. Arithm.* §272.

CHAPTER 8

8.1   Though it has a function similar to that of *Disqu. Arithm.* in number theory, *Th. mot.* is not quite as singular in the history of astronomy. There were several other contemporary efforts to present current knowledge in a systematic way, but Gauss was by far the most influential among them.

8.2   There is an extensive secondary literature about the priority question. It seems that the motivation, deduction, and systematic application of the method of least squares is more interesting than the problem of deciding who happened to discover, use, and publish it first.

8.3   It is curious that Gauss does not extensively discuss a question that fascinated and occupied him for many years.

8.4   See the relevant remarks in Brendel's essay in Vol. XI,2 of *G.W.*

8.5   In *Entwicklung der Mathematik im 19. Jahrhundert.*

8.6   There are strong indications that Gauss concerned himself with the agM as early as the early 1790's (see [Schlesinger] in Vol. X,2 of *G.W.*).

8.7   The lemniscate has the following shape.

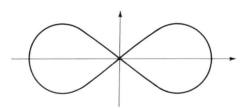

Analytically, it is characterized by

$$(x^2 + y^2)^2 = a^2(x^2 - y^2).$$

There is an instructive comparison of the properties of the trigonometric functions and the lemniscate in Markuschewitsch's essay in [Reichardt].

8.8   Gauss discovered, among others, the functions $\vartheta_{00}(\psi|x)$, $\vartheta_{01}(\psi|x)$, $\vartheta_{10}(\psi|x)$, $\vartheta_{11}(\psi|x)$ and many of the relations among them (cf. his posthumously published paper "Hundert neue Theoreme ..." in Vol. III of *G.W.*).

8.9   Letter #61 to Bessel, of March 30, 1828.

INTERCHAPTER VII

VII. 1   See [Schlesinger], p. 102, in Vol. X,2 of *G.W.* See also *G.W.* VIII, p. 103.

VII. 2   Gauss was familiar with the geometrical meaning of the substitutions $t + i$ and $1/t$.

VII. 3   It was mentioned above that Gauss was well aware of the group-theoretic structure of the classes in *Disqu. Arithm.*, but never explicitly discussed it.

VII. 4   In *Institutiones calculi integralis II* and in *Nova Acta Petropolitana* XII.

CHAPTER 9

9.1   As opposed to Descartes's theory according to which Earth would have to be oblate around the equator.

9.2   Gauss wrote several letters to obtain Epinal's points. He finally succeeded with Laplace's help but it turned out that most of Epinal's markers could not be found any more.

9.3   See letter #20 to Schumacher, of August 12, 1818.

9.4   ... Ich habe hier von einem Tag auf den andern auf Ihren Besuch gehofft und hoffe noch darauf, da ich unter 8 Tagen nicht von hier weg kann; es werden noch zwei Richtungen festgelegt werden müssen, die nach Wulsode, wo H. Müller jetzt ist, und die nach Kalbsloh, wohin er von da in einigen Tagen abgehen wird. Letztere ist deswegen nothwendig, weil die Möglichkeit des Durchhaus von Hauselberg nach Scharnhorst noch sehr problematisch ist, indem vielleicht das Terrain des Hassels selbst noch zu hoch ist. Von Kalbsloh aus ist diese Möglichkeit viel wahrscheinlicher, allein ich substituire doch ungern Kalbsloh für Hauselberg, da man am ersten Platze Wulfsode nicht sehen kann.

Wohin ich von hier gehe, ist noch ungewiss; ich hätte mich daher erst gern noch hier mit Ihnen besprochen. Nach meiner vorläufigen Rechnung liegt Wilsede 12,3 Meter über dem Fussboden der Göttinger Sternwarte. Haben Sie die Zenithdistanzen auf Michaelis gemessen, so können Sie nun schon vorläufig alles auf die Meeresfläche beziehen. Die Distanz Wilsede von Hamburg wird 42454 Meter ± seyn.

9.5   [Reichardt] contains a comprehensive map of Gauss's triangles (from Vol. IX of *G.W.*).

9.6   In a letter to Sartorius, reprinted in the Gauss-Bolyai correspondence.

9.7   ... Du willst nun mein aufrichtiges unverholenes Urtheil. Und dies ist, dass Dein Verfahren mir noch nicht Genüge leistet. Ich will versuchen, den Stein des Anstosses, den ich noch darin finde (und der auch wieder zu derselben Gruppe von Klippen gehört woran meine Versuche bisher scheiterten) mit so vieler Klarheit als mir möglich ist ans Licht zu ziehen. Ich habe zwar noch immer die Hoffnung, dass jene Klippen einst, und noch vor meinem Ende, eine Durchfahrt erlauben werden. Indess habe ich jetzt so manche andere Beschäftigungen vor der Hand ... (Letter #XXVII to Bolyai, of Nov. 25, 1804.)

9.8   Indications can be found in the correspondence with Gerling and, to a lesser degree, Schumacher.

9.9   See quote on p. 106.

9.10   In the section *Die Transcendentale Aesthetik* in Kant's *Kritik der reinen Vernunft.*

9.11   Letter #359 to Gerling, of June 23, 1846.

9.12   Letter #147 to Olbers, of April 28, 1817.

9.13   idem.

9.14   Bolyai's letter #XXXIII, of June 20, 1831.

9.15   Letter #XXXV to Bolyai, of March 6, 1832.

9.16   It is debatable how much new substance the Copenhagen prize essay contains, but it seems that, for the first time, Gauss used complex relations in it in order to formulate a condition when a mapping would be conformal.

9.17   . . . Diese Sätze führen dahin, die Theorie der krummen Flächen aus einem neuen Gesichtspunkt zu betrachten, wo sich der Untersuchung ein weites noch ganz unangebautes Feld öffnet. Wenn man die Flächen nicht als Grenzen von Körpern, sondern als Körper, deren eine Dimension verschwindet, und zugleich als biegsam, aber nicht als dehnbar betrachtet, so begreift man, dass zweierlei wesentlich verschiedene Relationen zu unterscheiden sind theils nemlich solche, die eine bestimmte Form der Fläche im Raume voraussetzen, theils solche, welche von den verschiedenen Formen, die die Fläche annehmen kann, unabhängig sind. Die letztern sind es, wovon hier die Rede ist: nach dem, was vorhin bemerkt ist, gehört dazu das Krümmungsmaass; man sieht abe leicht, dass eben dahin die Betrachtung der auf der Fläche construirten Figuren, ihrer Winkel, ihres Flächeninhalts und ihrer Totalkrümmung, die Verbindung der Punkte durch kürzeste Linien u. dgl. gehört. Alle solche Untersuchungen müssen davon ausgehen, dass die Natur der krummen Fläche an sich durch den Ausdruck eines unbestimmten Linearelements in der Form $\sqrt{(E\,dp^2 + 2F\,dp.dq + G\,dq^2)}$ gegeben ist. . . .

9.18   . . . In praktischer Rücksicht ist dies zwar ganz unwichtig, weil in der That bei den größten Dreiecken, die sich auf der Erde messen lassen, diese Ungleichheit in der Vertheilung unmerklich wird, aber die Würde der Wissenschaft erfordert doch, daß man die Natur dieser Ungleichheit klar begreife. . . .

9.19   Ohm's method is severely limited by the fact that it is useless for geometric problems. For more details, see [Bolza] in Vol. X,2 of *G.W.*

9.20   Gauss's was the first proof of Green's formula. Green's work, though not much later, was completely unknown outside England.

CHAPTER 10

10.1   Schumacher's circular, announcing and describing his two new journals, is included in the Gauss-Schumacher correspondence (June 1821). Schumacher also initiated a yearbook (*Jahrbuch*, an annual progress report) which the later correspondence frequently refers to.

10.2   See Gauss's correspondence with Schumacher, Olbers, and Bessel in the spring of 1824.

10.3   Tralles, a physicist.

10.4   Friedrich Wilhelm III, a thrifty and narrow-minded prince, was primarily interested in problems like the design of his soldiers' uniforms and similar, rather practical, questions.

10.5   . . . Darüber, dass er bei der Universität nicht angestellt wird, waren wir bereits alle einig. Da nun aber der Minister zur Herbeischaffung der Summe, welche noch an seiner Stellung fehlt, einen Titel haben muss, so hat er den Antrag an den König gemacht, den ich auch unterstützt habe, dass der Hr. Gauss ihm, dem Minister, in allem, was das mathematische Studium betrifft, rathgebend oder leitend für öffentliche Angelegenheiten oder Institute, als Observatorien, polytechnische Institute p.p. beistehe und sich unterzöge. Dies ist auch genehmigt, und der Minister hat dafür die Bewilligung auf 6 bis 700 Rthlr. erhalten, so dass von dieser Seite nun nichts mehr entgegensteht. Ausserdem würde noch eine billige Reise-und Versetzungskostenvergütung zu erlangen sein.

Was die Stellung betrifft, so glaube ich, dass neben der als Akademiker *sich keine ehrenvollere finden lässt*, und wenn der Hofrath Gauss sich mit dem Minister zu benehmen weiss, so bekommt er einen grossen Einfluss auf das ganze mathematische Unterrichtswesen des Staates, wo er also ein grosses Feld hat und ausserordentlich nützlich werden kann. Der Minister und die ersten Räthe werden ihm mit grossem Vertrauen entgegenkommen, alles übrige hängt von ihm selbst ab. Kommt es dazu, ein polytechnisches Institut zu bilden, wozu ich einen Plan entworfen habe, so würde er einen grossen Einfluss darauf üben, und dies ist zugleich eine Gelegenheit zu seiner Verbesserung. . . . (Letter of Nov. 28, 1824.)

10.6   See his obituary (*Gedächtnisrede*) of Dirichlet, reprinted in the second volume of his works.

10.7   Felix Klein liked to stress this aspect of the duties of the modern scientist.

10.8   Minna Gauss (-Waldeck), as well as her mother, seems to have favored Berlin.

10.9   Even then, more than 100 years ago, the scene resembled a sophisticated version of musical chairs. Mannheim, Düsseldorf, Tübingen, Berlin, and Seeberg were among the places involved.

10.10   There has not been any research into the history of the Waldeck family though the results of such investigations might be interesting. We know of a brother-in-law of Gauss, an officer who served and was killed in the Napoleonic wars.

10.11   For an example, see letter #435 to Schumacher, of July 9, 1845.

10.12   See note (6), Interchapter IV.

10.13   Letter #383 to Gerling, of December 30, 1852.

10.14   Not all of the letters to Gerling which concern this matter have actually been preserved. Several are contained in the recently published supplementary volume of the Gauss-Gerling correspondence (see Bibliography).

10.15   Letter #245 to Gerling, of March 19, 1836.

10.16   The letters that exist are reprinted in [Mack], a very valuable reference.

10.17   . . . Dein Brief machte auch den Kindern große Freude, Joseph fragte wol 10 mal, von Vater, wann kommt er wieder? auch Minna schien grossen Theil daran zu nehmen, die fragt aber besonders, bringt mir Vater auch etwas mit?

Könnte ich es Dir sagen lieber Junge, wie manchen traurigen Augenblick ich schon während Deiner Abwesenheit gehabt habe, auch abgerechnet Vater seine Krankheit. Carl—bester Carl, hast Du mich auch wahrlich lieb? ich fühle es, meine öftere Verstimmung muss Dich oft kränken; aber bei Gott es liegt nicht bei mir sie zu verbannen;—auch diese übertriebene Empfindlichkeit, ich kann nicht Herr ihrer werden, gewiss—o gewiss es (ist) jetzt Folge zu grosser Reizbarkeit der Nerven, aber es wird, es muss anders werden, denn bei Gott, ich fühle mich selbst höchst unglüchlich dadurch. Habe nur noch Geduld guter Junge und entzieh mir Deine Liebe darum nicht, es wird, es muss anders werden, mit diesem trüben Sinn mag ich nicht leben.

... Ein Glück, dass der Schluss der Ferien und Deine Mutter Dich wieder hier zu ·uns treiben, sonst fürchte ich giebt Dir H von Lindenau so viel zu schauen und zu horchen, dass Du darüber das Wiederkommen vergessen könntest. Glaube aber ja nicht, dass ich es Dir missgönne, ich freue mich so herzlich, wenn Du zufrieden bist, und das bist Du dort gewiss. O Gott im Himmel vermöcht ich Dich doch ganz so glücklich zu machen, wie Du es von mir erwartetest, bei Gott, mir fehlt nicht der Wille—aber die Kraft, möge der Himmel mir geben, dass die Kinder gute Menschen werden, so habe ich wenigstens einen Theil meiner Bestimmung erfüllt ....

10.18 Obgleich ich Himly fest in die Hand habe versprechen müssen, gar nicht zu schreiben, so kann ich es mir doch ohnmöglich versagen, an Dich, guter Carl, mein Versprechen zu brechen.—Wie unaussprechlich hat es mich beglückt, dass Deine Gesundheit leidlich ist, ach es ist ja jetzt mein höchstes Gut. Meine Gesundheit ist in den Hauptpunkten bedeutend besser, aber deswegen darfst Du nicht darauf rechnen mich im Äusseren eben verändert zu finden. Kummer und Krankheit haben mich zu tief heruntergebracht, als dass nicht erst eine längere Zeit dazu gehören sollte, bevor die tiefen Furchen wieder ausgeglichen sind.—Aber es wird auch kommen.—Was Du über Eugen schreibst, hat mich recht getröstet, Gott nimmt sich unserer noch an, wenigstens hab ich es mit dem heissesten Dank gegen Gott erkannt, dass er Dich ein Schiff finden liess, wie konnten wir es denn erwarten, dass sich gerade ein gleich Seegel fertiges fand.—Ach ja es ist das Letzte was Du für ihn thust.—Gott stehe ihm bei. Ist es mir doch als fühlte ich es ganz neu, es ist kein gestorbener, es ist ein verlorner Sohn Dass Du Dich mein bester Carl da etwas länger aufhalten musst, begreife ich wol, glaube nicht, dass Du erst den Tag Deiner Wiederkunft schreiben musst, der Tag Deiner Wiederkehr ist mir immer ein Festtag und wird Licht in diese dunkle Nacht bringen, die mich umgiebt. Nun noch eine Bitte, guter guter Carl, schlag sie nicht ab, richte Deine Reise so ein dass Du Wilhelm besuchst, die Ihssen hat geschrieben, so dringend darum gebeten, sie und iehr Mann wären überzeugt, dass es so gut auf ihn würken wurde, Du würdest Dich auch über ihn freun, Carl, Carl, thue es mir zum Trost, wir bedürfen ihn ja wol beide.—Joseph kann kein Hindernis geben, der ist erzogen und brav und gut,— aber Wilhelm soll erst noch werden ... Er hat an Dich geschrieben, ein Brief voll der heiligsten Betheurungen, wie es ein heisses Bestreben sein sollte Dir und mir Freude zu machen. Ihssens versichern, wie sie sich jetzt doppelt seiner annehmen wurden. Ach Karl könntest Du meine Bitte abschlagen? Gott ich bin so tief so tief gebeugt, ach ich flehe zu Dir, versage nicht was Du so leicht erfüllen kannst.—Bei Gott, ich will ja dann auch thun was in meinen Kräften ist, um mich aus dieser Kummer vollen Grabesnacht zu erheben. Ich kann nicht mehr bester Carl, Gott nehme Dich in seinen Schutz—Carl, Carl verwirf mein Flehen nicht.

10.19 Eine schwere Zeit für mein Haus sind alle die Monate gewesen, die seit meinem letzten Briefe an Sie verflossen sind. Ach wie lange und wie hart hat die arme Dulderin gedrückt werden mussen, bis ihr Herz brechen konnte. Endlich ist es gebrochen. Am 12. Abends ist sie von dem Jammer des Lebens geschieden, und heute hat die Erde ihre irdischen Überreste wieder aufgenommen. Meine beiden Töchter waren und sind mir eine wahre Stütze, meinen ältesten Sohn, welcher jetzt im Lüneburgischen eine Nachlese zu den vorjährigen Messungen hält, hoffe ich in ein paar Wochen hier zu sehen. Mein jüngster Sohn in Poppenhagen fängt eben an, sich von einer lebensgefährlichen Krankheit, die ihn vor etwa 6 Wochen befiel, zu erholen.

Wegen der Wiederbesetzung von Bohnenbergers Stelle hatte man um meinen Rath ersucht, ich hatte Gerling dazu vorgeschlagen, welchem auch unter sehr vortheilhaften Bedingungen die Vokation nach Tübingen zugekommen ist . . . (Letter to Olbers of Sept. 16, 1831).

10.20   For details, see the correspondence with Schumacher, mostly in the years 1828–1831.

10.21   Particularly in the correspondence with Olbers.

10.22   . . . Lebensfreude und Lebensmut waren schon lange von mir gewichen, und ich weiss nicht, ob sie je wiederkehren werden. Was mich so schwer drückt ist das Verhältnis zu dem Taugenichts in A(merika), der meinen Namen entehrt. Sie wissen, welche Nachricht ich vor 4 Monaten von ihm erhalten habe. Ich sehe, dass es wohl gut gewesen wäre, wenn ich ihm damals in dem Sinne geantwortet hatte, wie Sie rieten, um ihm sofort jede Erwartung abzuschneiden; aber ich vermochte nicht, ihm überhaupt zu antworten. Jetzt ist nun eine neue Epistel angekommen. Unschätzbar ware es, wenn ich Sie, mein teurer Freund, zur Stelle hätte, wie in so vielen anderen Rücksichten, so auch in der, dass Ihre bewährte Freundschaft und Ihr ungetrübter Blick meinem befangenen einen Stützpunkt geben könnte. Aber das Schicksal hat es nicht gewollt, mir diese Lebensfreude zu gewähren. Lassen Sie mich dann aber doch aus der Ferne, so gut es geht, Ihre Freundschaft in Anspruch nehmen, da aus naheliegenden Gründen hier niemand sich dazu eignet, mich darüber zu beraten. Ich lege den Brief selbst bei. Ich bitte Sie, liebster Gerling, mir Ihre Ansicht offen mitzuteilen, und enthalte mich, um Ihr Urteil ganz unbefangen zu erhalten, den Eindruck anzugeben, den er bei mir gemacht hat . . .

10.23   After the early 1820s, Gauss's letters to Bessel are notably more guarded than those to his other correspondents. A contributing reason may have been that Bessel was very persistent in his efforts to draw Gauss to Berlin and into Prussian service.

10.24   Such an attitude seems to have been quite characteristic for the time. Goethe reacted similarly, in a much more extreme way than Gauss did.

10.25   One of the visitors was the well-known Belgian scientist Quetelet who wrote a report of his visit which has been preserved.

10.26   Mostly in the correspondence with Schumacher.

10.27   Best known are Gauss's remarks about non-Euclidean geometry, another example is Gauss's comment that Abel's work saved him the trouble of publishing perhaps a third of his own results. According to Dedekind's biography of Riemann, Gauss made a similar statement about Riemann's dissertation.

Chapter 11

11.1   Gauss gave a balanced evaluation of the available candidates in his statement for the university administration, extolling Weber as an exceptionally gifted young researcher. Another candidate was Gerling.

11.2   See also Chapter 10, where we mention Gauss's work for the Hanoverian commission of weights and measures.

11.3   ... Wenn man durch s das Volumen der Flüssigkeit, durch h die Höhe ihres Schwerpunkts über einer beliebigen horizontalen Ebene, durch T den Inhalt desjenigen Theils der Oberfläche der Flüssigkeit, welche das Gefäss berührt, und durch U den Inhalt des anderen (freien) Theiles dieser Oberfläche bezeichnet: so ist im Zustande des Gleichgewichts das Aggregat

$$sh + (\alpha\alpha - 2\beta\beta)T + \alpha\alpha U$$

ein Minimum, wo α und β gewisse Constanten bedeuten, welche von dem Verhältniss der Schwere zu der Intensität der Molecularanziehung der Theile der Flüssigkeit gegen einander und der Theile des Gefässes gegen die Flüssigkeit abhängen. Wir sehen hier also, als die Frucht einer schwierigen und subtilen Untersuchung einen Ausdruck für das Gesetz des Gleichgewichts hervorgehen, der, selbst dem gemeinen Verstande begreiflich, die Vermittlung des Conflicts zwischen den verschiedenen hier ins Spiel tretenden Kräften klar vor Augen legt. Wäre die Schwere die einzige wirkende Kraft, so würde beim Gleichgewicht der Schwerpunkt der ganzen Flüssigkeit so tief wie möglich liegen, also h ein Minimum sein müssen. Setzt man hingegen die Schwere und die Anziehung des Gefässes ganz bei Seite, so dass bloss die gegenseitige Anziehung der Theile der Flüssigkeit selbst in Betracht kommt, so muss diese eine sphärische Gestalt annehmen, also $T + U$ ein Minimum sein. Wäre endlich weder Schwere noch gegenseitige Anziehung der Flüssigkeitstheile vorhanden, so würde die Flüssigkeit sich über die ganze Oberfläche des Gefässes verbreiten, also T ein Maximum oder $-T$ ein Minimum sein müssen. Man findet es begreiflich, dass beim Zusammenwirken der drei Kräfte ein aus jenen drei Grössen-Zusammengesetztes ein Kleinstes werden soll, wiewohl sich von selbst versteht, dass die eigentliche feste Begründung jenes Lehrsatzes nur auf die vollständigen strengen mathematischen Schlussreihen gestützt werden kann, die von der Natur der Molecularanziehung wesentlich abhängig sind.

11.4   [Bolza] gives more details of the history of the expression. It seems to have made its first appearance in scholastic philosophy.

11.5   Gauss's problem is the mutual attraction of two arbitrary spherical ellipsoids. The transformation into confocal spherical ellipsoids is possible with the help of MacLaurin's theorem.

11.6   See note (20), chap. 9.

11.7   This may have been one of the factors in the increasing (temporary) alienation between Gauss and Humboldt in the early 1830s.

11.8   Lamont was later recognized as one of the world's leading researchers in magnetism. For his curious and interesting career, see the entry *Lamont* in one of the older editions of *Encyclopaedia Britannica*.

11.9   ... Dass die unbedeutenden Versuche, die Ich vor 5 Jahren bei Ihnen zu machen das Vergnügen hatte, mich der Beschäftigung mit dem Magnetismus zugewandt hätten, kann ich zwar nicht eigentlich sagen, denn in der That ist mein Verlangen danach so alt, wie meine Beschäftigung mit den exacten Wissenschaften überhaupt, also weit über 40 Jahr; allein ich habe den Fehler, dass ich erst dann recht eifrig mich mit einer Sache beschäftigen mag, wenn mir die Mittel zu einem rechten Eindringen zu Gebote stehen und daran fehlte as früher. Das freundschaftliche Verhältniss, in welchem ich zu unserm trefflichen Weber stehe, seine ungemein grosse Gefälligkeit, alle Hülfsmittel des Physikalischen Cabinets zu meiner Disposition zu stellen und mich mit seinem eignen Reichthum an praktischen Ideen zu unterstützen, machte mir die ersten Schritte erst möglich, und den ersten Impuls dazu haben doch wieder Sie gegeben, durch einen Brief an Weber, worin Sie (Ende 1831) der unter Ihren Auspicien errichteten Anstalten für Beobachtung der täglichen Variation erwähnen . . . (Letter of June 13, 1833).

11.10   Though Gauss's work is on much better foundations than that of his predecessors, it is not without gaps. Gauss specifically induced the existence of a minimum from the existence of a lower bound. This is the essence of Dirichlet's principle; Gauss, of course, did not prove it, nor did he see the need for a proof.

11.11   Humboldt made such a comment in a letter to Bessel.

11.12   See letter #333 to Olbers, of Nov. 1, 1837, and also later letters to Olbers.

11.13   K. Schwarzschild, "Zur Elektrodynamik I–III," *Nachrichten der Gesellschaft der Wissenschaften zu Göttingen, Math.-Phys. Klasse.* 1903.

11.14   The fragment is contained in Vol. V of *G.W.*

INTERCHAPTER VIII

VIII.1   Quoted after [Dunnington].

VIII.2   Letter #347 to Olbers, of March 8, 1839.

VIII.3   See letters to Schumacher #225, of March 15, 1836, and #226, of March 21, 1836. See also letter #439, of Sept. 22, 1845.

VIII.4   See Gauss's memoranda (*pro memoria*) about the Hanover survey in Vol. IV of *G.W.*

VIII.5   There is again a certain similarity to Goethe and his reactions.

VIII.6   For more details, see W. Harich, *Jean Pauls Revolutionsdichtung*, Berlin 1974.

CHAPTER 12

12.1   Letter #171 to Schumacher, of January 28, 1831.

12.2   The question became moot in 1866 when the Kingdom of Hanover was dissolved and incorporated into Prussia. Weber, originally from Saxony, another victim of Prussia's vigorous expansionism, developed into a glowing Hanoverian patriot and never forgave the Prussians their crime.

12.3   During the 19th century, universities were acknowledged and accepted as havens of liberalism in Germany. Though King Ernest August was no more reactionary than other contemporary German princes, his strong action was quite exceptional.

12.4   Auf Ihren Brief vom 8. mein theuerster Freud, der aber erst heute in meine Hände gekommen ist, eile ich Ihnen sogleich zu antworten, dass der mich betreffende Zeitungsartikel (. . .), insofern eine der vielen Unwahrheiten ist, die jetzt die öffentlichen Blätter füllen, als ich mich wirklich bisher gegen Niemand übr das geäussert habe, was ich zu thun oder nicht zu thun willens sei. Ich wünsche und hoffe, dass die Universität als Corpus, sich in die politischen Wirrnisse nicht mischen möge. Indessen wissen Sie, dass zwei mir sehr nahe stehende Personen insofern hineingezogen sind, als sie sich haben bewegen lassen, die bekannte Vorstellung mit zu unterzeichnen. Die deshalb beim Universitätsgericht eingeleitete Untersuchung betrifft indessen nur, wenn ich recht berichtet bin, die unbefugte Verbreitung, und an dieser haben die zwei angedeuteten gewiss auch nicht den entferntesten Theil gehabt. Ich kann daher nicht glauben, dass jene Unterzeichnung für sie unangenehme Folgen haben werde, und so lange diese zwei kräftigen Magnete unverrückt und unbeschädigt sind, behält Göttingen für mich viel mehr Reiz als Paris. Ob aber, wenn je Umstände eintreten sollten, die mir das Leben in Göttingen verbitterten, ich Paris anderen Orten vorziehen würde, ist etwas was jetzt hier nicht in Frage zu kommen braucht.... (Letter #263 to Schumacher, of December 13, 1837).

12.5   So Gauss did much more to retain his friends in Göttingen than is commonly assumed. The only step—albeit a critical one—he was reluctant to take and in fact did not take was to voice his concern publicly. This would have been against all his convictions. In 1838, apparently to heal the rift, Gauss was elected rector of the university, but he turned the office down.

12.6   Letter #221 (Gauss to Gerling) in the correspondence with Gerling (June 28, 1833).

12.7   There are several references in the correspondences with A. v. Humboldt and Schumacher.

INTERCHAPTER IX

IX.1   See Note (2), Chapter 8.

IX.2   This is Gauss's own assessment.

IX.3   See Note (7), Interchapter VI.

IX.4   Es ist sehr merkwürdig, dass die freien Bewegungen, wenn sie mit den nothwendigen Bedingungen nicht bestehen können, von der Natur gerade auf dieselbe Art modificirt werden, wie der rechnende Mathematiker, nach der Methode der Kleinsten Quadrate, Erfahrungen ausgleicht, die sich auf unter einander durch nothwendige Abhängigkeit verknüpfte Grössen beziehen. Diese Analogie liesse sich noch weiter verfolgen, was jedoch gegenwärtig nicht zu meiner Absicht gehört. (*G.W.* V, p.28)

IX.5   Aber nicht bloss unsere Armuth documentirt eine solche Art zu urtheilen, sondern zugleich eine kleinliche, engherzige und träge Denkungsart, eine Disposition, immer den Lohn jeder Kraftäusserung ängstlich zu calculiren, einen Kaltsinn und eine Gefühllosigkeit gegen das Grosse und den Menschen Ehrende. Man kann es sich leider nicht verheelen, dass man eine solche Denkungsart in unserm Zeitalter sehr verbreitet findet, und es ist wohl völlig gewiss, dass gerade diese Denkart mit dem Unglück, was in den letzten Zeiten so viele Staaten betroffen hat, in einem sehr genauen Zusammenhange steht; verstehen Sie mich recht, ich spreche nicht von dem so häufigen Mangel an Sinn für die Wissenschaften an sich, sondern von der Quelle, woraus derselbe fliesst, von der Tendenz, überall zuerst nach dem Vortheil zu fragen, und alles auf physisches Wohlsein zu beziehen, von der Gleichgültigkeit gegen grosse Ideen, von der Abneigung gegen Kraftanstrengung bloss aus reinem Enthusiasmus für eine Sache an sich: ich meine, dass solche Charakterzüge, wenn sie sehr vorherrschend sind, einen starken Ausschlag bei den Katastrophen, die wir erlebt haben, gegeben haben können. (*G.W.*, Vol. XII, p. 192.)

IX.6   See, e.g., letter #321 to Olbers, of Jan. 20, 1835, or a letter to the philosopher Fries, dated May 11, 1841. There are various interesting remarks in the correspondence with Schumacher. Concerning Hegel, see letters #333 of Jan. 23, 1842, #334, of Jan. 25, 1842, and #335, of Feb. 2, 1842. Concerning Wolff, see the letter #412, of Nov. 1, 1844.

IX.7   Cf. Note (6) above. Vol. XII of *G.W.* contains a letter from Weber to Fries, transmitting Gauss's high appreciation of Fries's work.

Chapter 13

13.1   According to his own estimate, Gauss had to manipulate more than one million numbers for the geodetic survey alone. He also had to perform enormous calculations in order to reduce his astronomical data.

13.2   See, as an example, in Vol. II of *G.W.* his review of J. Ch. Burckhardt's tables of divisors.

13.3   It is actually not possible for us to tell whether Gauss found his estimates inductively or not. There are no clues how he could have derived them but it seems unlikely that they are not based on some theoretical consideration.

13.4   See Gauss's letter to Encke of March 11, 1851.

13.5   Gauss made a few remarks about the "undulation theory" of light but he never involved himself in this question. One should keep in mind that Gauss knew Fraunhofer from his trip to Munich in 1816. See also letter #411 to Schumacher, where Gauss discusses a paper which Herschel had submitted to one of Schumacher's journals.

13.6   More about the initial contacts between Gauss and Repsold in [Schaefer] in Vol. XI,2 of *G.W.*, pp. 152ff.

13.7   Like [Klinkerfues].

13.8   See [Schaefer] for a discussion of this point, pp. 182ff.

Chapter 14

14.1   There are reports by Cantor, the historian of mathematics, Dedekind, Jacobi, and others.

14.2   Among them the 8 queens problem (See correspondence with Schumacher or Vol. XII, p. 18ff. of *G.W.*). Curiously enough, Gauss used the complex plane to treat the problem.

14.3   . . . Indessen, wenn Sie, wie ich hoffe, ihn auch jetzt wieder annehmen, so bin ich doch in Sorge, was einmal künftig aus ihm werden soll. Die Geschichte mit dem Barometer, die vielleicht nicht die einzige ist, lässt mich fürchten, dass es ihm an Geschick für praktische Astronomie fehlt. Diese aber u der Lehrstand sind ja jetzt in Europa fast das Einzige, wie ein Mathematiker, der keine eigenen Mittel hat, seine Subsistenz sichern kann. Sie wissen, wie unsere Akademien jetzt beschaffen sind, und nur wenn er etwas ganz Eminentes leistete, wäre einige Hoffnung, dass er einmahl in einer Akademie eine Versorgung finde, und selbst dann ist 99 gegen 1 zu wetten, dass das nicht glückt. Ob er nun einmahl ein Professorenamt bekleiden kann, weiss ich nicht; Sie werden dies besser beurtheilen können. Fürchten Sie aber, dass er auch dazu sich nicht eignet, so weiss ich nicht, ob er nicht am besten thäte, irgend einen andern Beruf zu erwählen, z.B. als Militär oder sonst, und dann seine Musse nach Gefallen der Mathematik widmete. In der That, wenn man einmal einen Brotberuf dabei nöthig hat, so ist es ziemlich einerlei, welcher es ist, ob man Anfängern das abc der Wissenschaften vorträgt, oder Schuhe macht. Die Frage bleibt eigentlich nur, bei welcher Arbeit man die meiste und sorgenfreieste Zeit übrig behält . . .

14.4   Letter #374 to Schumacher, of Oct. 12, 1843.

14.5   Letter #342 to Schumacher, of Sept. 16, 1842.

14.6   According to Rudio, Eisenstein went to Göttingen in 1844, his first trip after completing his education at the *Gymnasium* in Berlin. (See Eisenstein, Collected Papers, Vol. II)

14.7 Our best source is [Dedekind b].

14.8 Dedekind, *Werke* Bd. II. Cf. Bibliography [Dedekind b].

14.9 See the list of courses given by Gauss in [Dunnington].

14.10 Gauss was a good and effective champion of Lobachevski's work in Germany.

14.11 ... Ich habe die Veranlassung gehabt, das Werkchen von Lobatschefski (Geometrische Untersuchungen zur Theorie der Parallellinie. Berlin 1840, bei G. Funcke. 4 Bogen stark) wieder durchzusehen. Es enthält die Grundzüge derjenigen Geometrie, die Statt finden müßte, und streng consequent Statt finden könnte, wenn die Euclidische nicht die wahre ist. Ein gewisser Schweikardt nannte eine solche Geometrie Astralgeometrie, Lobatschefsky imaginäre Geometrie. Sie wissen, dass ich schon seit 54 Jahren (seit 1792) dieselbe Überzeugung habe (mit einer gewissen späteren Erweiterung, deren ich hier nicht erwähnen will). Materiell für mich Neues habe ich also im Lobatschefsky'schen Werke nicht gefunden, aber die Entwickelung ist auf anderem Wege gemacht, als ich selbst eingeschlagen habe, und zwar von Lobatschefsky auf eine meisterhafte Art in ächt geometrischem Geiste. Ich glaube Sie auf das Buch aufmerksam machen zu müssen, welches Ihnen gewiss ganz exquisiten Genuss gewähren wird.

14.12 Gauss owned several volumes of Pushkin and was very fond of *Boris Godunov*. See Gauss's letters ##283 and 308 to Schumacher (of Aug. 17, 1839, and Aug. 8, 1840).

14.13 See letter #218 to Schumacher, of Sept. 27, 1835.

14.14 See Gauss's report in Vol. IV of *G.W.*

14.15 Miss Waldeck was quite wealthy, and Gauss took care to keep the accounts apart. Minna Gauss made a separate will (with stipulations when Eugen would be entitled to obtain his share), and one can see the development of the family finances by looking at her and his will.

14.16 See letter #412 to Schumacher, of Nov. 1, 1844.

14.17 —so everything appeared to go well.

14.18 Letter #386, of April 21, 1853, in the correspondence with Gerling, and subsequent letters.

14.19 ... Es sind die Stellen, die sich auf Unsterblichkeit beziehen. Ich kann Ihnen jetzt nicht sagen, woher die Zusammenstellung. Aber ich finde sie doch alle nicht so schlagend und zusammenhängend. Überhaupt, lieber College, ich glaube, Sie sind viel bibelgläubiger als ich und Sie sind viel glücklicher als ich. Ich muß sagen, wenn ich so öfters in früheren Zeiten Leute in niederen Ständen, simple Handwerker, gesehen, die so recht von Herzen glauben konnten: ich habe sie immer beneidet. Sagen Sie mir doch, wie fängt man dieß an? ... Haben Sie vielleicht das Glück gehabt, einen gläubigen Vater zu haben, oder eine Mutter? *Nachrichten*, Akad. Wiss. Göttingen, Phil.-hist. Klasse. 1975. No. 6

CHAPTER 15

15.1 In his letter of Dec. 7, 1853, Gauss congratulated Alexander von Humboldt on having reached Newton's age, namely the sum of 30766 days—an either meaningless or enormous compliment.

15.2 See Dedekind's short biography of Riemann, in Riemann's collected works.

15.3 Jacobi's letter to his brother, of Sept. 1, 1849.

15.4 It is the topic of several letters in the Gauss-Gerling correspondence in the course of the year 1853.

15.5 As a student of Dirichlet, Dedekind might be called a mathematical "grandson" or, better, "grandnephew" of Gauss.

APPENDIX B

B.1 Several of the essays in Vls. X,2 and XI,2 were first published separately, as publications of the Göttingen Royal Society; in several cases, there are substantial and interesting changes between the two versions.

B.2 A history of the development of number theory in the 19th century does, unfortunately, not yet exist.

B.3 In 'Gauss zum Gedächtnis'.

B.4 These two papers are quite different from the rest of the essays. They are less historically oriented but contain a wealth of material.

Bibliography

Part A

[Dunnington] contains a chronological listing of Gauss's works. We confine ourselves to the most important references. For more information, one should consult Gauss's *Collected Works*, specifically the essays in Vols. X,2 and XI,2.

Carl Friedrich Gauss, *Werke I–XII*, edited and published by Königliche Gesellschaft der Wissenschaften zu Göttingen between 1863 and 1933. In 1973, this edition was reprinted by Georg Olms Verlag, Hildesheim/New York.

The following entry concerns a paper of Gauss that was not included in *G.W.* (see p. 78 for a discussion).

Ozhigova, E. P.: C. F. Gauss, Übersicht über die Gründe der Constructibilität des Siebenzehneckes. Istoriko-mat. Issledovaniya 21, 1976.

CORRESPONDENCE

*Briefwechsel zwischen Gauss und Bessel,*G. F. Auwers, ed. Leipzig 1880. Reprint Hildesheim/New York 1975.

Briefwechsel zwischen C. F. Gauss und Wolfgang Bolyai, F. Schmidt and P. Stäckel, eds. Leipzig 1899. Reprint New York/London 1972.

Briefwechsel zwischen Carl Friedrich Gauss und Christian Ludwig Gerling, C. Schaefer, ed. Berlin 1927. Reprint Hildesheim/New York 1975.

Christian Ludwig Gerling and Carl Friedrich Gauss. Sechzig bisher unveröffentlichte Briefe, T. Gerardy ed. Göttingen 1964.

Cinque lettres de Sophie Germain à C. F. Gauss, B. Boncampagni, ed. Berlin 1880.

Briefe zwischen A. v. Humboldt und Gauss, K. Bruhns, ed. Leipzig 1877.

Briefwechsel zwischen Alexander von Humboldt und Carl Friedrich Gauss, K.-R. Biermann, ed. Berlin 1977.

Briefe von C. F. Gauss an B. Nicolai, W. Valentiner, ed. Karlsruhe 1877.

Briefwechsel zwischen Olbers und Gauss, C. Schilling, ed. Berlin 1900/1905. Reprint Hildesheim/New York 1976.

Briefwechsel zwischen C. F. Gauss und H. C. Schumacher, C. A. F. Peters, ed. Altona 1860–1865. Reprint Hildesheim/New York 1975.

Nachträge zum Briefwechsel zwischen Carl Friedrich Gauss und Heinrich Christian Schumacher, T. Gerardy, ed. Göttingen 1969.

Other important sources are Vol. XII of *G.W.*, Abhandlungen 71 (1955) of Bayr. Akademie der Wissenschaften, Math.-Naturwiss. Abteilung, and Mitteilungen der Gauss-Gesellschaft e.V. Göttingen. Eisenstein's correspondence with Gauss can be found in Vol. 2 of the former's collected papers. Other letters, to a variety of correspondents, have appeared in numerous places; [Dunnington] and Biermann's edition of the correspondence with A. v. Humboldt contain many references.

Part B

Bieberbach, L. *Carl Friedrich Gauss, ein deutsches Gelehrtenleben*. Berlin 1938.

Biedermann, K. *Deutschland im 18. Jahrhundert*. 1854. Reprint Aalen 1969.

Biermann, K.-R. (a) DDR-Schrifttum über C. F. Gauss. Mitteilungen der Math. Ges. der DDR 4, 1974.

Biermann, K.-R. (b) *Die Mathematik und ihre Dozenten an der Berliner Universität 1810–1920*. Berlin 1973.

Biermann, K.-R. (c) Martin Bartels—eine Schlüsselfigur in der Geschichte der nicht-euklidischen Geometrie? Mitteilungen der Leopoldina Halle 1975.

Black, M. *The Nature of Mathematics*. New York 1952.

Bourbaki, N. *Eléments d'histoire de mathématique*. Paris 1960

Bruford, K. *Germany in the 18th Century*. Cambridge 1935.

Crowe, M. J. *A History of Vector Analysis*. South Bend. 1967.

Dedekind, R. (a) Vorlesungen über Zahlentheorie von P. G. L. Dirichlet. Braunschweig 1863^1, 1871^2, $1879/80^3$, 1894^4.

Dedekind, R. (b) *Gauss in seiner Vorlesung über die Methode der kleinsten Quadrate*. Berlin 1901.

Dieudonné, J. *L'oeuvre mathématique de C. F. Gauss*. Conférence du Palais de la Découverte D.79 Paris 1962.

Dombrowski, P. Differentialgeometrie . . . (Lecture, Braunschweig 1977).

Du Moulin-Eckart, R. *Geschichte der deutschen Universitäten*. Stuttgart 1929.

Dunnington, G. W. *C. F. Gauss, Titan of Science*. New York, 1955.

Edwards, H. M., *Fermat's Last Theorem*. New York 1977.

Fries, J. Die Geschichte der Philosophie . . . Halle 1837–1840.

Gerth, H. *Die sozialgeschichtliche Lage der bürgerlichen Intellegenz um die Wende des 18. Jahrhunderts*. Frankfurt 1935.

Goldstine, H. H. *A History of Numerical Analysis*, New York 1977.

Hadamard, J. *The Psychology of Invention in the Mathematical Field*. Princeton 1945.

Hänselmann, L. *Carl Friedrich Gauss, zwölf Kapitel aus seinem Leben*. Leipzig 1878.

Hall, T. *Gauss*. Cambridge 1970.

Jones, E. Das Problem Paul Morphy. Psychoanalytische Bewegung III, 1931.

Klein, F. *Entwicklung der Mathematik im 19. Jahrhundert I.* Berlin 1925.

Klinkerfues, W. *Theoretische Astronomie.* Braunschweig 1871.

Koenigsberger, L. *Zur Erinnerung an Jacob Friedrich Fries.* Sitzungsberichte der Heidelberger Akademie der Wiss., Math.-naturw. Klasse. Heidelberg 1911.

Kronecker, L. *Vorlesungen über Zahlentheorie I.* Leipzig 1901. Reprinted Berlin 1978.

Kronecker, L. *Gesammelte Werke II.* Berlin 1897.

Leiste, C. *Die Arithmetik und Algebra.* Wolfenbüttel 1790.

Mack, H. *C. F. Gauss und die Seinen.* Braunschweig 1927.

Mitgau, J. H. *Familienschicksal und soziale Rangordnung.* Leipzig 1928.

Möbius, P. J. *Über die Anlage zur Mathematik.* Leipzig 1900.

Möser, J. J. *Sämtliche Werke.* Oldenburg 1944 –

Moritz, K. F., *Anton Reiser.* Berlin 1785 ff.

Paulsen, F. (a) *Das deutsche Bildungswesen.* Leipzig 1906.

Paulsen, F. (b) *Geschichte des gelehrten Unterrichts auf den deutschen Schulen und Universitäten vom Ausgang des Mittelalters bis zur Gegenwart.* Leipzig 1885.

Reich, K. *Carl Friedrich Gauss 1777/1977.* Bonn–Bad Godesberg 1977.

Reichardt, H. (Ed.) *C. F. Gauss Gedenkband anlässlich des 100. Todestages am 25. Februar 1955. Leipzig 1957.*

Reichardt, H. *Gauss und die nicht-euklidische Geometrie.* Leipzig 1976.

Riecke, E. *Wilhelm Weber.* Göttingen 1892.

Rosen, V. H. On mathematical 'illumination' and the mathematical thought process. The Psychoanalytic Study of the Child 8, 1953.

Scharlau, W. and Opolka, H. *Von Fermat bis Minkowski—Eine Vorlesung über Zahlentheorie und ihre Entwicklung.* Heidelberg 1980.

Schmidt, A. *Nachrichten von Büchern und Menschen I.* Frankfurt 1971.

Sheynin, O. B. Gauss and the Theory of Errors. Archive for History of Exact Sciences 20, 1979.

Smend, R. Die Göttinger Gesellschaft der Wissenschaften in *Festschrift zur Feier des zweihundertjährigen Bestehens der Akademie der Wissenschaften in Göttingen I.* Berlin 1951.

Wagner, R. Gespräche mit Carl Friedrich Gauss in den letzten Monaten seines Lebens. Nachrichten Akad. Wiss. Göttingen, Philolog. histor. Klasse. 1975, No. 6.

Weil, A. Two lectures on number theory, past and present. Enseign. Math. XX, 1974.

Weber, H. *Lehrbuch der Algebra.* Leipzig 1894–1908.

Whitehead, A. N. *Science and the modern world.* London 1925.

Wolff, C. *Anfangsgründe aller mathematischen Wissenschaften.* Frankfurt 1750–1757.

Worbs, E. *Carl Friedrich Gauss: Ein Lebensbild.* Leipzig 1955.

Added in proof: Carl Friedrich Gauss: A Bibliography. Wilmington (Del.) 1981.

Index

N.B. This index does not contain any references to the Preface, the Appendices, or the Bibliography.